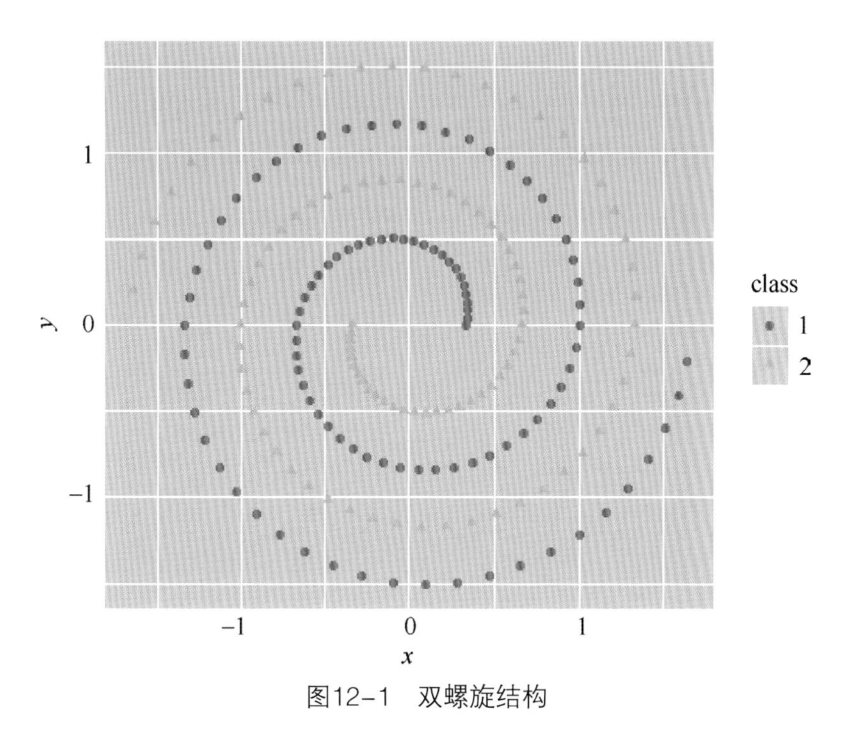

图12-1　双螺旋结构

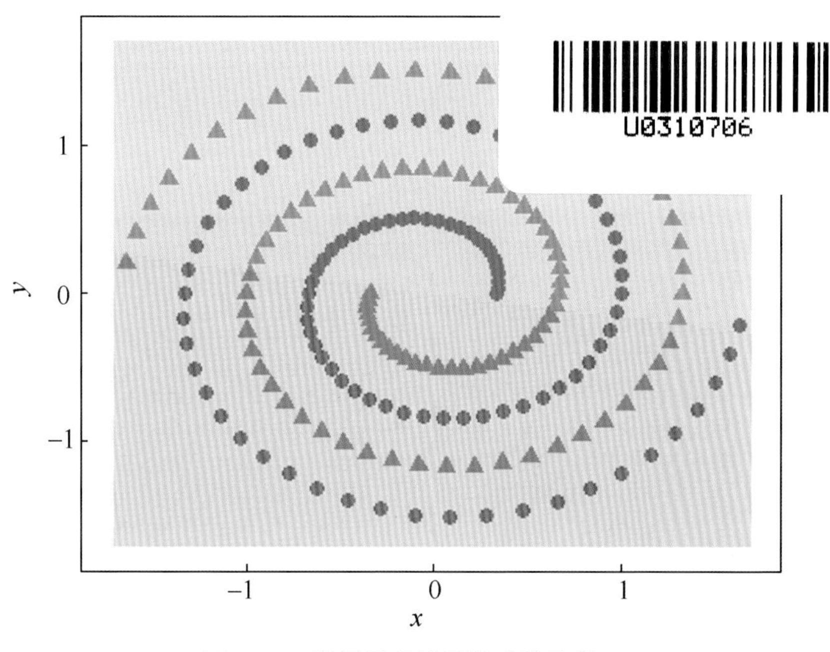

图12-3　逻辑回归模型的决策边界

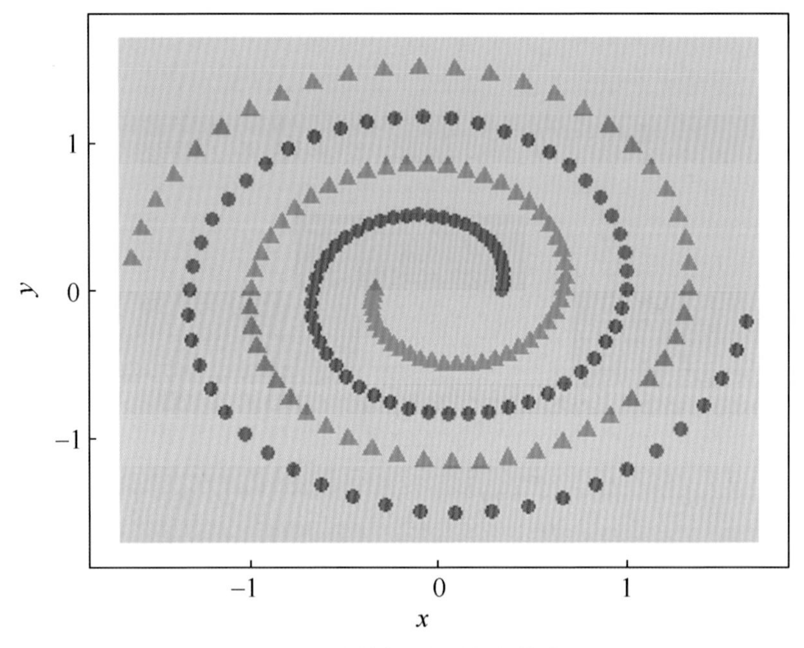

图12-4　决策树模型的决策边界

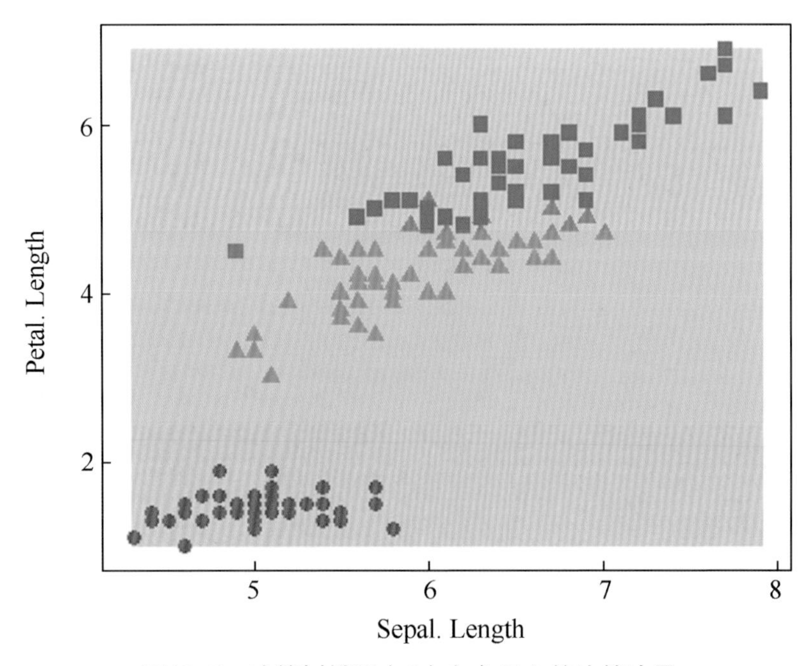

图12-5　决策树模型在2个自变量上的决策边界

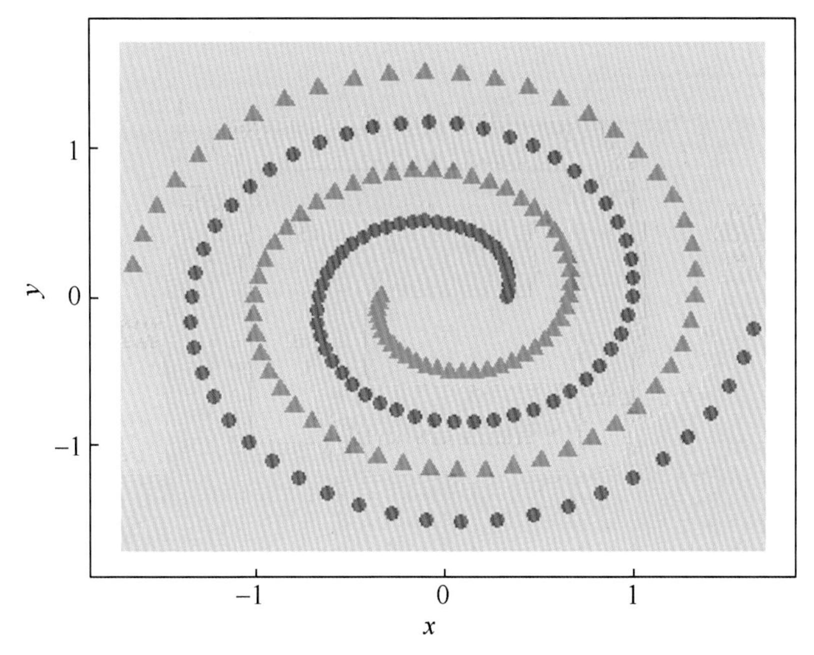

图12-6　随机森林模型的决策边界

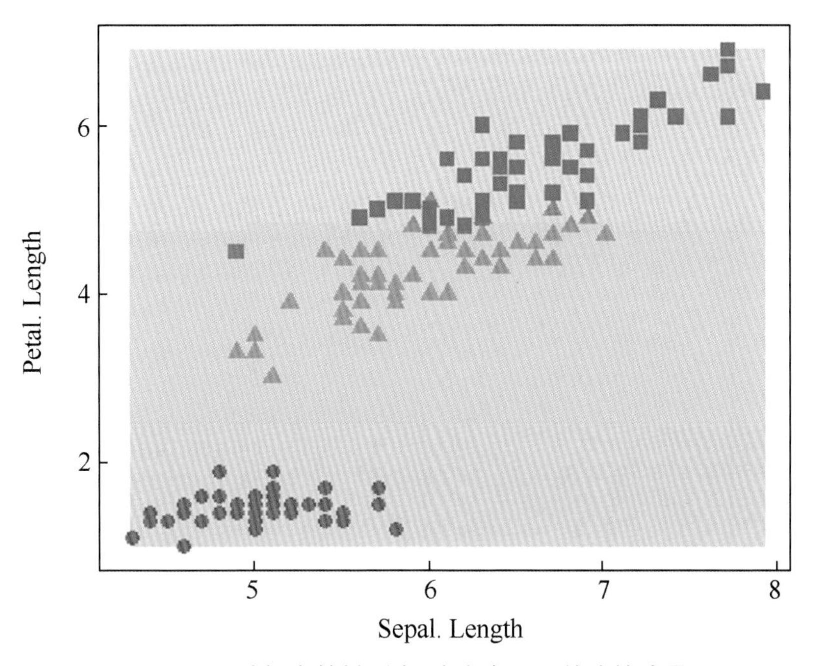

图12-7　随机森林模型在2个自变量上的决策边界

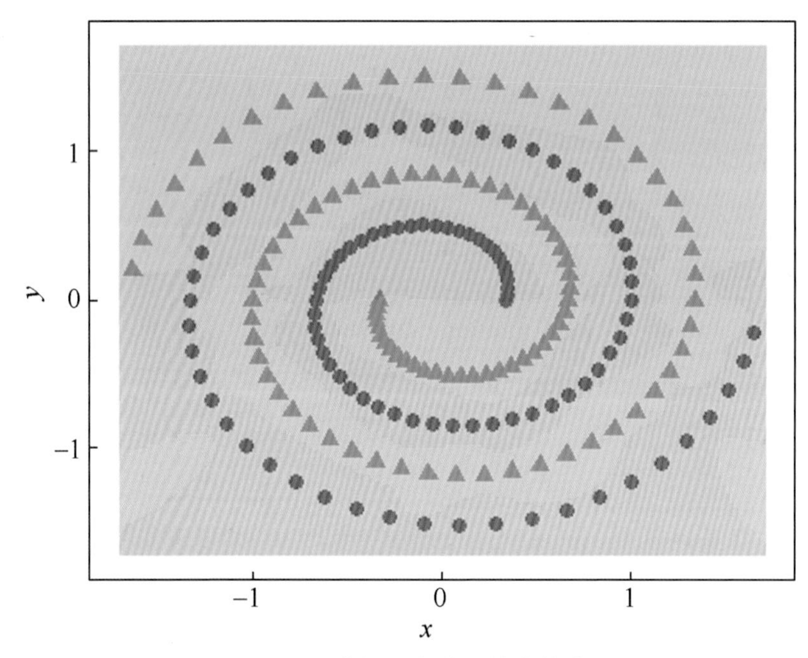

图12-10　神经网络模型的决策边界

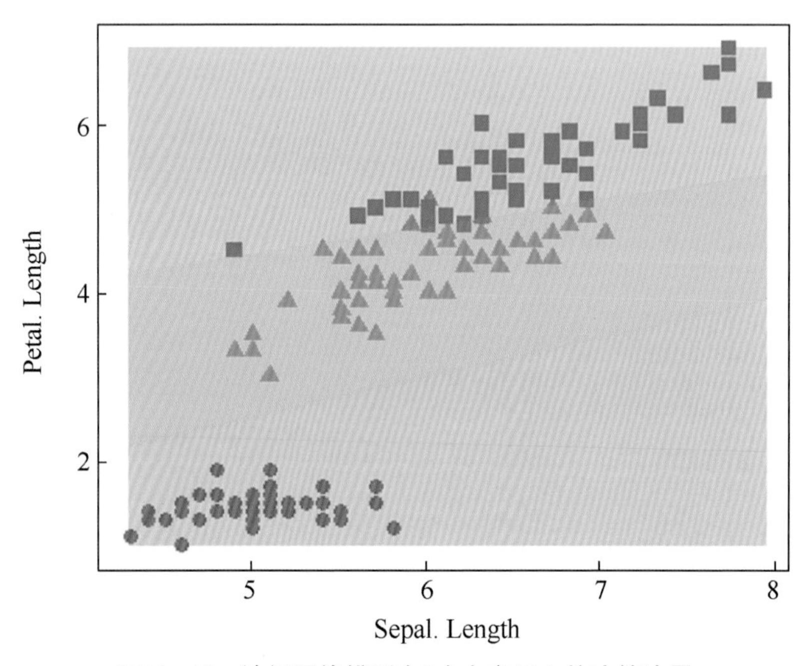

图12-11　神经网络模型在2个自变量上的决策边界

高等学校大数据技术与应用规划教材

Hadoop 大数据分析

高永彬　钱亮宏　方志军　编著

中国铁道出版社有限公司
CHINA RAILWAY PUBLISHING HOUSE CO., LTD.

内 容 简 介

本书从 Hadoop 的原理和使用出发，在重点介绍 Hadoop 生态系统的重要组件 HDFS、MapReduce、YARN、Hive 和 Spark 的同时，注重大数据分析能力的全面提高。

本书共分 13 章，主要内容包括 Hadoop 简介、HDFS 文件系统、YARN 资源管理、MapReduce 计算框架、Hive 简介、Hive 数据定义、Hive 数据操作、Hive 数据查询、Spark 简介、Spark 大数据处理、Spark 机器学习流程、Spark 有监督学习模型和 Spark 无监督学习模型。

本书内容丰富、体系新颖、结构合理、文字精练，适合作为普通高等院校信息类专业 Hadoop 大数据分析课程的教材，也可以作为数据科学行业相关从业人员的自学教材。

图书在版编目（CIP）数据

Hadoop 大数据分析/高永彬，钱亮宏，方志军编著. —北京：中国铁道出版社有限公司，2019.7（2020.8 重印）
高等学校大数据技术与应用规划教材
ISBN 978-7-113-25919-8

Ⅰ.①H… Ⅱ.①高…②钱…③方… Ⅲ.①数据处理软件-高等学校-教材 Ⅳ.①TP274

中国版本图书馆 CIP 数据核字（2019）第 119302 号

书　　名：Hadoop 大数据分析
作　　者：高永彬　钱亮宏　方志军

策　　划：曹莉群　　　　编辑部电话：010-51873090
责任编辑：包　宁
封面设计：穆　丽
责任校对：张玉华
责任印制：樊启鹏

出版发行：中国铁道出版社有限公司（100054，北京市西城区右安门西街 8 号）
网　　址：http://www.tdpress.com/51eds/
印　　刷：北京柏力行彩印有限公司
版　　次：2019 年 7 月第 1 版　2020 年 8 月第 2 次印刷
开　　本：787 mm×1 092 mm 1/16　插页：2　印张：11.5　字数：287 千
书　　号：ISBN 978-7-113-25919-8
定　　价：38.00 元

前　言

随着信息技术的普及和应用，各行各业产生了大量的数据，人们持续不断地探索处理这些数据的方法，以期最大限度地从中挖掘有用信息。面对如潮水般不断增加的数据，人们不再满足于数据的查询和统计分析，而是期望从数据中提取信息或者知识为决策服务。数据挖掘技术突破数据分析技术的种种局限，结合统计学、数据库、机器学习等技术解决从数据中发现新的信息并辅助决策这一难题，是正在飞速发展的前沿学科。近年来，随着教育部“新工科”建设的不断推进，大数据技术受到广泛关注。数据挖掘作为大数据技术的重要实现手段，能够挖掘数据的关联规则，实现数据的分类、聚类、异常检测和时间序列分析等，解决商务管理、生产控制、市场分析、工程设计和科学探索等各行各业中的数据分析与信息挖掘问题。

Hadoop是一系列分布式存储和计算软件，由Doug Cutting创建，能够支持互联网数据量级别的系统。狭义的Hadoop项目仅包含Hadoop Common、HDFS、YARN和MapReduce 4个组件。广义的Hadoop项目还包含了其他一些衍生性的项目组件，它们或多或少依赖以上4个核心组件，如数据存储依赖于HDFS、作业调度和资源管理依赖YARN，同时它们还解决了一些特定领域的问题。常用的包括Spark、HBase、Hive、Sqoop、Oozie、Impala、Hue、Pig等。

截至2019年1月，共有283所高校获批“数据科学与大数据技术”专业，其中985及211高校占比13%。目前国内大数据人才缺口更是达到百万级。由于其开源性、易用性和强大的数据分析能力，Hadoop已成为世界范围内应用最广泛的数据科学工具和语言之一。目前，Hadoop大数据分析与挖掘逐渐成为高校信息类专业的必修课，同时，作为面向各专业的通识课也广受欢迎。

本书作为立足于应用型本科数据科学与大数据教学的Hadoop核心课教材，具有如下特色：

（1）内容安排合理且全面，从Hadoop的安装配置、分布式数据处理、分布式数据仓库到分布式机器学习，循序渐进，深入浅出。

（2）难度适中，适用于本科中高年级的核心课教材，仅需掌握Python基本编程和Linux基本操作就可以学习本书，对Java编程及数学和算法知识不作为必要基础。

（3）理论与案例相结合，理论与实践相结合，包含了泰坦尼克号乘客生存分析、航班准点数据处理、鸢尾花数据建模等实践案例。

本书主要内容分为以下3部分：

第1部分：Hadoop核心基础，包括第1～4章。第1章为Hadoop简介，包括Hadoop的相关背景、基本概念、安装、配置和运行等。第2章为HDFS文件系统，包括HDFS架构、文件库和常用操作等。第3章为YARN资源管理，包括YARN架构、调度策略和常用操作等。第 4 章为 MapReduce 计算框架，包括各 MapReduce 原理、流程、词频统计和数据连接的实现等。

第 2 部分：Hive 数据仓库，包括第 5～8 章。第 5 章为 Hive 简介，包括 Hive 的相关背景、基本概念、安装、配置和运行等。第 6 章为 Hive 数据定义，包括数据库操作、数据表操作、数据格式、外部表和分区表等。第 7 章为数据操作，包括数据导入、数据插入和数据导出等。第 8 章为 Hive 数据查询，包括基本查询、数据聚合和数据连接等。

第 3 部分：Spark 数据分析，包括第 9～13 章。第 9 章为 Spark 简介，包括 Spark 的相关背景、基本概念、安装、配置和运行等。第 10 章为 Spark 大数据处理，包括大数据的选择、聚合、引用、筛选、连接和变形等。第 11 章为 Spark 机器学习流程，包括数据探索、划分、填充、特征选择、建模调优和测试评估等。第 12 章为 Spark 有监督学习模型，包括线性、决策树、随机森林、神经网络和协同过滤等。第 13 章为 Spark 无监督学习模型，包括 k 均值聚类、主成分分析和关联分析模型等。

本书例子中的所有数据都可在 GitHub 上公开下载，地址为 https://github.com/yepdata/hadoop_textbook。

本书由高永彬、钱亮宏和方志军编著。具体分工如下：高永彬编写第 1～4 章；方志军编写第 5～8 章，钱亮宏编写第 9～13 章。全书由范磊和许华根主审。同时感谢戴仁月、严娟和刘敏对本书的贡献。

由于编者水平有限，加之时间仓促，书中难免存在疏漏和不足之处，敬请老师和同学批评指正。

编　者

2019 年 5 月

目　录

第 1 部分　Hadoop 核心基础

第 2 部分　Hive 数据仓库

Hadoop 核心基础

第 1 章

Hadoop 简介 «‹

1.1 Hadoop 产生背景

随着互联网的爆炸式发展，全球的数据量正以指数级增长。据工信部表示，我国数据量正以每年 50%的速度增长，预计到 2020 年中国数据总量将达到 8.8 ZB（1 ZB = 1024 EB，1 EB = 1024 PB），而世界数据总量将达到 40 ZB。据 IBM 称，整个人类文明所获得的全部数据中，有 90%是过去两年内产生的。随着数据体量的增加，大数据时代悄然来临。

全球范围内数以亿计的人们每天与互联网接触，而这些线上行为也产生了丰富的数据轨迹。互联网公司能够捕捉到每一次与用户的交互，包括线上搜索、网页点击、停留时间和达成交易。数据不仅仅是存储量上的增长，同时还更加丰富和多样化。产生了大量图片、视频、文本、语音和传感器信号等不同于传统的能用行和列表示的结构化数据。

随着数据量的增加，采用单台高性能服务器已无法处理和分析全量数据。即使单台高性能服务器能够满足需要，其硬件成本投入与计算能力并不线性相关，即计算能力提升一倍时硬件成本投入通常需要提升远超一倍。而基于多台商用服务器的分布式集群则能够较好地处理和分析大数据。所谓商用服务器，指的并不是已淘汰的服务器或家用 PC，而是相对于定制化大型机而言的企业级通用服务器，硬件成本相比于定制化大型机要低很多。

分布式系统并不只是并行的存储和处理数据这么简单，还需要考虑许多复杂的系统问题。第一个问题是应对硬件故障以实现高可用性。随着服务器（在集群中又称节点）数量的增加，集群中有硬件故障的概率将非常高。为避免数据丢失，最常用的方法是将数据复制多份，放在不同的服务器上，这样即使一台服务器出现故障，仍能够获取数据的其他副本。Hadoop 分布式文件系统（Hadoop Distributed File System，HDFS）即采用了这一策略。另一个问题是数据的合并方式。多数数据处理和分析过程都需要某种形式的数据合并，即输出数据的某一条记录并不仅仅由输入数据的某一条记录决定，而是由输入数据的多条记录所决定。MapReduce 计算框架将数据处理的每一阶段拆分成 Map 和 Reduce 阶段，用于抽象所有的数据处理和分析过程，具体细节将在后续章节介绍。Hadoop 并不是第一个分布式数据存储和处理平台，甚至任何开发人员都可以从头自行搭建分布式系统。Hadoop 的魅力在于，分布式系统中那些琐碎的后勤工作都已经实现（如数据的副本存储在哪个节点或数据处理任务分配给哪个节点），而开发人员只需要去实现业务逻辑即可。

简言之，Hadoop 是一个可靠且可扩展的分布式大数据存储和处理平台，可以在商用服务器上运行，完全开源，因此性价比很高。

1.2 Hadoop 简要历史

Hadoop 由 Doug Cutting 创建，他同时也是文本搜索库 Apache Lucene 项目的创建人。Hadoop 原先是开源网页搜索引擎 Apache Nutch 项目的一部分，而 Nutch 同时又是 Lucene 的一部分。

Nutch 创建于 2002 年，然而该项目的创建人意识到他们原先的系统架构无法扩展到数以亿计的网页。而就在 2003 年 10 月，谷歌发表了一篇描述谷歌分布式文件系统（Google File System，GFS）架构的论文。2004 年，Nutch 的开发人员实现了该论文的思想，并命名为 Nutch 分布式文件系统（Nutch Distributed File System，NDFS）。

2004 年 12 月，谷歌又发表了一篇介绍 MapReduce 计算框架的论文。2005 年，Nutch 的开发人员同样实现了 MapReduce，并将 Nutch 中的所有算法迁移到了 MapReduce 计算框架和 NDFS 文件系统中。

Nutch 中 NDFS 和 MapReduce 的实现完全可以应用于搜索引擎以外的领域。因此，2006 年 2 月，这部分从 Nutch 中独立成了 Lucene 下面的一个子项目，称为 Hadoop。与此同时，Doug Cutting 加入了雅虎，带领一个团队全身心投入，使得 Hadoop 成为能够支持互联网数据量级别的系统。2006 年 4 月，Hadoop 0.1 版本正式发布。

2008 年 1 月，Hadoop 称为 Apache 顶级项目，不再作为 Lucene 的子项目存在，证明了自己的成功。与此同时，许多雅虎以外的公司（如脸书等）也开始使用 Hadoop 并贡献代码。2008 年 4 月，Hadoop 用 910 个节点在 209 s 内排序了 1 TB 数据，创造了世界纪录，2007 年的纪录是 297 s。2009 年 4 月，雅虎宣布 Hadoop 用 62 s 排序了 1 TB 数据。

1.3 Hadoop 生态系统组件

最初的 Hadoop 是由 HDFS 文件系统和 MapReduce 计算框架组成的。MapReduce 本质上是一个批处理系统，并不适合做交互式分析。这意味着用户提交一个查询后，无法在几秒内得到结果。多数在 MapReduce 中运行的查询需要几分钟甚至更长时间，因此比较适合做离线数据处理。

正是由于 MapReduce 应用场景的局限性，在 Hadoop 诞生后的几年时间里，衍生出了许多其他项目组件，这些项目组件都可以理解为 Hadoop 生态系统中的项目，同时 Hadoop 这个词从广义上来说也都包含了这些项目组件。这些项目大多也是 Apache 软件基金会的项目。例如，HBase 就是一个能快速响应查询结果的分布式存储系统，适用于需要快速读/写的数据应用。

Hadoop 生态系统组件不断壮大的过程中，YARN（Yet Another Resource Negotiator）的引入功不可没。它是一个集群资源管理系统，使得任意分布式应用（而不仅仅是 MapReduce 程序）可以在 Hadoop 集群中运行。

狭义的 Hadoop 项目仅包含如下 4 个组件：

- Hadoop Common：Hadoop 核心组件，其他所有组件都依赖该核心组件，为 Hadoop

的其他组件提供了一些常用工具，主要包括系统配置工具、序列化机制和 Hadoop 抽象文件系统等。

- HDFS：Hadoop 分布式文件系统，为 Hadoop 的其他组件提供高吞吐量的数据访问、管理文件的存储位置和副本情况。
- YARN：Hadoop 作业调度和资源管理系统，即用户提交作业的排队方式、分配资源（主要为 CPU 核心和内存）的数量和调度具体执行任务的节点。
- MapReduce：基于 YARN 的并行处理大数据的计算框架，用户只需要实现 Map 和 Reduce 阶段，其他后勤工作由框架完成。

广义的 Hadoop 项目还包含了其他一些衍生性的项目组件，它们或多或少依赖以上 4 个核心组件，如数据存储依赖 HDFS、作业调度和资源管理依赖 YARN，同时它们还解决了一些特定领域的问题。Hadoop 生态系统的项目组件如图 1-1 所示。

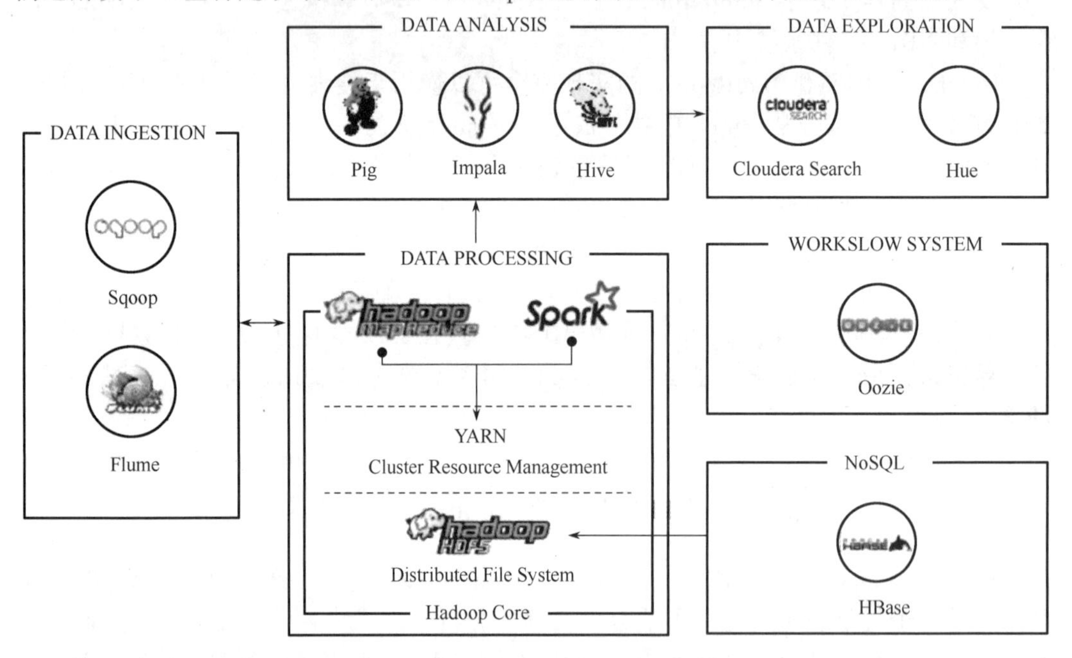

图 1-1 Hadoop 生态系统的项目组件

- Spark：大数据快速处理和分析的通用引擎，编程模型简单，支持多种应用，包括 ETL（extraction, transformation & loading）、机器学习、流处理和图计算等。
- HBase：非 SQL 型数据库，用于存储结构化的大数据表，由于查询不用通过 YARN 做资源调度，因此响应速度很快。
- Hive：SQL 型数据仓库，用于做数据的批处理，吞吐量高但响应速度较慢。
- Sqoop：数据导入/导出工具，用于从传统关系型数据库（如 MySQL 和 Oracle 等）导入/导出数据到 HDFS 或 Hive。
- Oozie：工作流调度系统，用于定义一系列工作流程以及执行路径，并按特定频率执行或由特定事件触发。
- Impala：SQL 型数据仓库，对 Hive 做了较大优化，响应速度上做了较大提升，主要由 Cloudera 公司主导开发。

- Hue：Hadoop 平台的网页接口，使得分析人员能够通过网页界面提交作业并查看结果，提升了用户体验。
- Pig：数据处理工具，用简单的脚本语言做复杂的变换、聚合和分析操作。

1.4 Hadoop 版本和商用支持

自从 2006 年 4 月 Hadoop 0.1 版本正式发布，Hadoop 生态系统中的项目持续不断演进，开源社区一直保持活跃，几次重大的版本升级和增加的特性如下。

- 2011 年 11 月，Hadoop 1.0 版本发布，意味着 Hadoop 已经完全适用于企业生产环境，包括了许多企业安全策略方面的支持。
- 2012 年 5 月，Hadoop 2.0 版本发布，增加了 HDFS 多命名节点的机制，即可以配置多个命名节点，则单个命名节点的故障不会引发整个 HDFS 文件系统无法访问；另外，引入了 YARN 作为作业调度和资源管理系统。
- 2016 年 9 月，Hadoop 3.0 版本发布，支持 2 个以上命名节点，HDFS 支持纠删码。

同时，随着 Hadoop 生态系统的不断丰富，大数据软件市场也充满了机遇。Hadoop 虽然已经大大简化了用户搭建分布式大数据系统的流程，但仍然需要较高的技术能力，尤其是应对突发平台故障的能力。另外，Hadoop 生态系统中的各项目组件存在多种版本，不同版本之间的兼容性和稳定性对于企业用户而言至关重要。因此，一些提供 Hadoop 技术服务的公司应运而生，它们会基于开源 Hadoop 生态系统各项目组件，经过严格的版本兼容性和稳定性测试，打包并发布稳定的发行版，并在此基础上提供管理工具、技术支持和咨询服务。其中，最具有代表性的是 Cloudera 和 Hortonworks。

Cloudera 公司发布的 Hadoop 发行版简称为 CDH（Cloudera Distribution of Hadoop），也是世界范围内应用最广泛的 Hadoop 发行版，本书所有例子都将在 CDH 上运行。截至本书撰稿时，CDH 的最新版本为 6.1，其对应的 Hadoop 生态系统项目组件的版本为：Hadoop 3.0.0、Hive 2.1.1、HBase 2.1.1、Spark 2.4、Sqoop 1.4.7、Oozie 5.0.0，以及其他组件。

类似的，Hortonworks 公司发布的 Hadoop 发行版简称为 HDP（Hortonworks Data Platform）。截至本书撰稿时，HDP 的最新版本为 3.1，其对应的 Hadoop 生态系统项目组件的版本如图 1-2 所示。

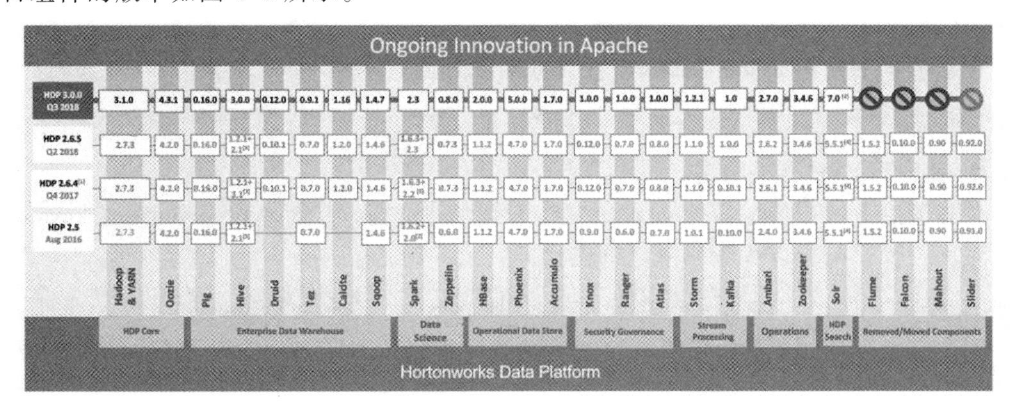

图 1-2　Hadoop 生态系统项目组件的版本

2018 年 10 月，Cloudera 和 Hortonworks 公司正式合并，意味着 Hadoop 的标准将更加统一。

1.5 Hadoop 的基础环境配置

搭建分布式 Hadoop 涉及 3 个节点，其中，主机名为 master 的节点为主节点，运行 HDFS 命名节点、YARN 资源管理等服务，同时也运行从节点的 HDFS 数据节点、YARN 节点管理等服务；主机名为 slave1 和 slave2 的节点为从节点，运行 HDFS 数据节点、YARN 节点管理等服务。

以下步骤中，主机名和 IP 地址以真实环境为准。以下每个步骤都尽可能注明执行位置为主节点、从节点或所有节点。

1. 关闭防火墙（所有节点）

CentOS 的防火墙服务 firewalld 用于限制特定端口的外部访问。Hadoop 包含了多种服务，使用到了不同端口，因此需要将所有节点的防火墙关闭。

```
sudo service firewalld stop
```

2. 主机名映射配置（所有节点）

配置集群中各组成节点的主机名和 IP 地址的映射，打开文件/etc/hosts。

```
vim /etc/hosts
```

在文件末尾添加如下内容，以实际的主机名和 IP 地址为准。

```
192.168.95.128 master192.168.95.129 slave1192.168.95.130 slave2
```

3. 时钟同步（所有节点）

启动时钟同步服务。

使用 systemctl enable 命令配置时钟同步服务在系统启动时自动启动。

```
systemctl enable ntpd
```

使用 systemctl start 命令启动时钟同步服务。

```
systemctl start ntpd
```

在从节点上使用 ntpdate 命令与主节点同步时钟。

```
ntpdate master
```

4. 免密码 SSH 连接配置

在主节点上使用 ssh-keygen 命令生成主节点的密钥，并加入到 authorized_keys 文件末尾作为授权的可以免密码登录的主机。

```
ssh-keygen -t dsa -P '' -f ~/.ssh/id_dsacat ~/.ssh/id_dsa.pub >> ~/.ssh/authorized_keys
```

在主节点上使用 ssh 命令测试与主节点的免密码 SSH 连接。如果不需要输入密码，则表示配置成功。

```
ssh master
```

在从节点上使用 scp 命令从主节点复制主节点的密钥到从节点，并加入到文件 authorized_keys 末尾作为授权的可以免密码登录的主机。

```
scp master:~/.ssh/id_dsa.pub ~/master_dsa.pubcat ~/master_ das.pub >> ~/.ssh/authorized_keys
```

在主节点上使用 ssh 命令测试与从节点的免密码 SSH 连接。如果不需要输入密码，则表示配置成功。

```
ssh slave1
ssh slave2
```

1.6 Hadoop 的安装

本节以在 CentOS 7.5 操作系统上搭建分布式 CDH 6.1 为例（该版本于 2018 年 12 月发布），搭建其他版本的过程类似。整个安装过程要求保持互联网连接。

1. 配置软件库（主节点）

CentOS 属于 RHEL 型操作系统，使用 yum 作为包管理工具，依赖于软件库来安装软件。

配置 CDH 6.1 的软件库，创建并打开软件库文件/etc/yum.repos.d/cloudera-cdh6.repo。

```
vim /etc/yum.repos.d/cloudera-cdh6.repo
```

使其看起来如下所示。

```
[cloudera-cdh6]
# Packages for Cloudera's Distribution for Hadoop, Version 6, on RedHat or CentOS 7 x86_64
name=Cloudera's Distribution for Hadoop, Version 6
baseurl=https://archive.cloudera.com/cdh6/6.1.0/redhat7/yum/
gpgkey=https://archive.cloudera.com/cdh6/6.1.0/redhat7/
yum/RPM-GPG-KEY-cloudera gpgcheck = 0
```

使用 rpm 命令导入软件库 GPG 密钥。

```
sudo rpm --import https://archive.cloudera.com/cdh6/6.1.0/redhat7/ yum/RPM-GPG-KEY-cloudera
```

2. 安装 Java 环境（主节点）

Hadoop 的运行和开发需要 Java 环境，即 Java Development Kit（JDK）。从 Java 官方网站（http://www.oracle.com/technetwork/java/javase/downloads/java-archive-javase8- 2177648.html）下载最新的 Java 版本，本节以 Java SE Development Kit 8u192 为例（见图 1-3）。选择接受授权条款，即 Accept License Agreement，并选择下载 Linux x64 的 tar.gz 版本，即 jdk-8u192-linux-x64.tar.gz。需要注意的是，Hadoop 暂时还不支持 JDK 9。

使用 tar 命令将 jdk-8u162-linux-x64.tar.gz 文件解压缩到文件夹/usr/java。

```
sudo tar zxf jdk-8u192-linux-x64.tar.gz -C /usr/java/
```

在主节点上使用 scp 命令从主节点复制 Java 路径到从节点的相同路径。

```
scp -r /usr/java/jdk-8u192-linux-x64 slave1:/usr/java/s
cp -r /usr/java/jdk-8u192-linux-x64 slave2:/usr/java/
```

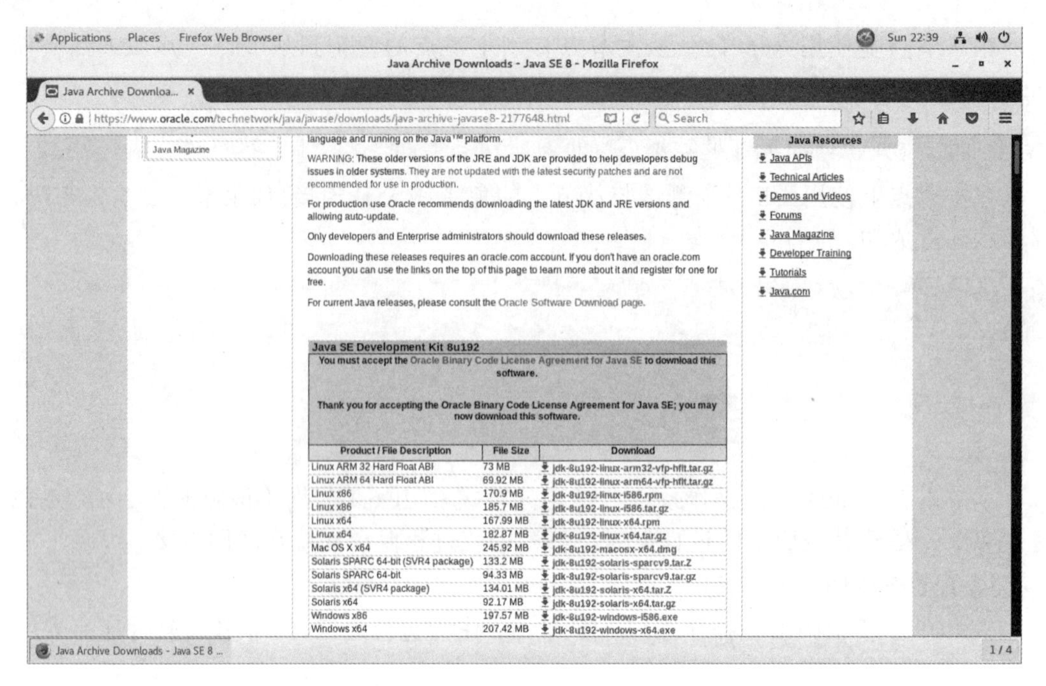

图 1-3　Java 官方网站

3. 安装关系型数据库 MySQL（主节点）

Hadoop 中的一些组件（如 Hive、Sqoop、Hue 或 Oozie）需要关系型数据库存储一些辅助信息才能正常运行，如 Hive 需要将元数据（metadata）存储在关系型数据库中。本节以 MySQL 为例。

使用 wget 命令下载 MySQL 的软件库。

```
sudo wget http://repo.mysql.com/mysql-community-release-el7-5.noarch.rpm
```

使用 rpm 命令导入 MySQL 的软件库。

```
sudo rpm -ivh mysql-community-release-el7-5.noarch.rpm
```

使用 yum 命令安装 MySQL 软件包。

```
sudo yum install mysql-server
```

打开 MySQL 配置文件。

```
vim /etc/my.cnf
```

修改文件，使其看起来如下所示。

```
[mysqld]
datadir=/var/lib/mysql
socket=/var/lib/mysql/mysql.sock
transaction-isolation = READ-COMMITTED
# Disabling symbolic-links is recommended to prevent assorted security risks;
# to do so, uncomment this line:
symbolic-links = 0

key_buffer_size = 32M
```

```
max_allowed_packet = 32M
thread_stack = 256K
thread_cache_size = 64
query_cache_limit = 8M
query_cache_size = 64M
query_cache_type = 1

max_connections = 550
#expire_logs_days = 10
#max_binlog_size = 100M

#log_bin should be on a disk with enough free space.
#Replace '/var/lib/mysql/mysql_binary_log' with an appropriate path for your
#system and chown the specified folder to the mysql user.
log_bin=/var/lib/mysql/mysql_binary_log

#In later versions of MySQL, if you enable the binary log and do not set
#a server_id, MySQL will not start. The server_id must be unique within
#the replicating group.
server_id=1

binlog_format = mixed

read_buffer_size = 2M
read_rnd_buffer_size = 16M
sort_buffer_size = 8M
join_buffer_size = 8M

# InnoDB settings
innodb_file_per_table = 1
innodb_flush_log_at_trx_commit = 2
innodb_log_buffer_size = 64M
innodb_buffer_pool_size = 1G
innodb_thread_concurrency = 8
innodb_flush_method = O_DIRECT
innodb_log_file_size = 512M

[mysqld_safe]
log-error=/var/log/mysqld.log
pid-file=/var/run/mysqld/mysqld.pid

sql_mode=STRICT_ALL_TABLES
```

使用 systemctl enable 命令配置 MySQL 服务在系统启动时自动启动。

```
sudo systemctl enable mysqld
```

使用 systemctl start 命令启动 MySQL 服务。

```
sudo systemctl start mysqld
```

执行安装脚本/usr/bin/mysql_secure_installation 配置 MySQL 中 root 用户的密码以及其他安全选项，这里将 root 用户密码设置为 123456。其他提示则按照如下示例输入。

```
sudo /usr/bin/mysql_secure_installation
[...]
```

```
Enter current password for root (enter for none):
OK, successfully used password, movimng on...
[...]
Set root password? [Y/n] y
New password: 123456
Re-enter new password: 123456
Password updated successfully!
Reloading privimlege tables..
 ... Success!
[...]
Remove anonymous users? [Y/n] y
 ... Success!
[...]
Disallow root login remotely? [Y/n] n
 ... Skipping.
[...]
Remove test database and access to it? [Y/n] y
 - Dropping test database...
[...]
Reload privimlege tables now? [Y/n] y
 ... Success!
[...]
```

4. 安装关系型数据库 MySQL JDBC 驱动器（主节点）

使用 rpm 命令安装 MySQL JDBC 驱动器。

```
sudo yum install mysql-connector-java
```

5. 安装 Hadoop 核心组件（所有节点）

使用 yum install 命令在线安装 Hadoop 核心组件，包括：

- HDFS 命令节点服务：hadoop-hdfs-namenode.x86_64;
- HDFS 数据节点服务：hadoop-hdfs-datanode.x86_64;
- YARN 资源管理器服务：hadoop-yarn-resourcemanager.x86_64;
- YARN 节点管理器服务：hadoop-yarn-nodemanager.x86_64。

在主节点上安装所有 4 个服务。

```
sudo yum install hadoop-hdfs-namenode.x86_64
sudo yum install hadoop-hdfs-datanode.x86_64
sudo yum install hadoop-yarn-resourcemanager.x86_64
sudo yum install hadoop-yarn-nodemanager.x86_64
```

在从节点上仅安装 HDFS 数据节点和 YARN 节点管理器服务。

```
sudo yum install hadoop-hdfs-datanode.x86_64
sudo yum install hadoop-yarn-nodemanager.x86_64
```

1.7 Hadoop 的配置

1. 使用默认配置（主节点）

使用 cp 命令将默认配置文件夹复制一份到文件夹/etc/hadoop/conf.my_cluster，后续配置都基于该默认配置修改。

```
sudo cp -r /etc/hadoop/conf.empty /etc/hadoop/conf.my_cluster
```

使用 alternatives 命令设置配置的优先级，并显示当前配置。

```
sudo alternatives --install /etc/hadoop/conf hadoop-conf /etc/hadoop/ conf.my_cluster 50
sudo alternatives --set hadoop-conf /etc/hadoop/conf.my_cluster
sudo alternatives --display hadoop-conf
hadoop-conf - status is manual.
link currently points to /etc/hadoop/conf.my_cluster
/etc/hadoop/conf.empty - priority 10
/etc/hadoop/conf.impala - priority 5
/etc/hadoop/conf.my_cluster - priority 50
Current 'best' version is /etc/hadoop/conf.my_cluster.
```

可以看出，文件夹/etc/hadoop/conf.my_cluster 对应的配置优先级最高（50），为当前使用的配置。

2. Hadoop 的核心配置（主节点）

打开 Hadoop 核心配置文件/etc/hadoop/conf.my_cluster/core-site.xml，加入的配置项包括：

- fs.defaultFS：分布式文件系统的默认地址和端口，如果配置分布式多节点集群，则需要将这里的 localhost 替换成命名节点所在的主机名；
- hadoop.proxyuser.hive.hosts：用户 hive 作为超级用户可以发起连接的主机名，后续章节运行 Hive 时需要；
- hadoop.proxyuser.hive.groups：用户 hive 作为超级用户可以模拟的用户组，后续章节运行 Hive 时需要。

该文件配置部分（即<configuration>标签内）看起来如下所示。

```
vim /etc/hadoop/conf.my_cluster/core-site.xml
<configuration>
    <property>
        <name>fs.defaultFS</name>
        <value>hdfs://master:8020</value>
    </property>
    <property>
        <name>hadoop.proxyuser.hive.hosts</name>
        <value>*</value>
    </property>
    <property>
        <name>hadoop.proxyuser.hive.groups</name>
        <value>*</value>
    </property>
</configuration>
```

3. HDFS 的配置（主节点）

打开 HDFS 配置文件/etc/hadoop/conf.my_cluster/hdfs-site.xml，加入的配置项包括：

- dfs.permissions.superusergroup：HDFS 超级用户组；
- dfs.replication：HDFS 中每个文件块的副本份数，这里设置为 3 份；如果 Hadoop 为单节点伪分布式，则将副本数设为 1 份。

该文件配置部分（即<configuration>标签内）看起来如下所示。

```
vim /etc/hadoop/conf.my_cluster/hdfs-site.xml
<configuration>
    <property>
        <name>dfs.permissions.superusergroup</name>
        <value>hadoop</value>
    </property>
    <property>
        <name>dfs.replication</name>
        <value>3</value>
    </property>
</configuration>
```

4. MapReduce 的配置（主节点）

打开 MapReduce 配置文件/etc/hadoop/conf.my_cluster/mapred-site.xml，加入的配置项包括：

- mapreduce.framework.name：MapReduce 作业所用的资源调度框架为 YARN；
- yarn.app.mapreduce.am.resource.mb：MapReduce 作业的应用主程序所占用内存；
- mapreduce.map.memory.mb：MapReduce 作业的 Map 任务所占用内存；
- mapreduce.reduce.memory.mb：MapReduce 作业的 Reduce 任务所占用内存。

该文件配置部分（即<configuration>标签内）看起来如下所示。

```
vim /etc/hadoop/conf.my_cluster/mapred-site.xml
<configuration>
    <property>
        <name>mapreduce.framework.name</name>
        <value>yarn</value>
    </property>
    <property>
        <name>yarn.app.mapreduce.am.resource.mb</name>
        <value>512</value>
    </property>
    <property>
        <name>mapreduce.map.memory.mb</name>
        <value>512</value>
    </property>
    <property>
        <name>mapreduce.reduce.memory.mb</name>
        <value>512</value>
    </property>
</configuration>
```

5．YARN 的配置（主节点）

打开 YARN 配置文件/etc/hadoop/conf.my_cluster/yarn-site.xml，加入的配置项包括：

- yarn.nodemanager.resource.memory-mb：YARN 节点管理器所能分配的内存；
- yarn.nodemanager.resource.cpu-vcores：YARN 节点管理器所能分配的 CPU 核数；
- yarn.scheduler.minimum-allocation-mb：YARN 容器所占用的最小内存；
- yarn.scheduler.maximum-allocation-mb：YARN 容器所占用的最大内存；
- yarn.scheduler.increment-allocation-mb：YARN 容器内存的最小增加量；
- yarn.application.classpath：YARN 应用的 JAR 包路径；
- yarn.nodemanager.aux-servimces：YARN 节点管理器的辅助服务，这里添加 MapReduce 的 Shuffle 操作；
- yarn.nodemanager.env-whitelist：YARN 节点管理器的环境变量白名单；
- yarn.resourcemanager.hostname：YARN 资源管理器的主机名。

该文件配置部分（即<configuration>标签内）看起来如下所示。

```
vim /etc/hadoop/conf.my_cluster/yarn-site.xml
<configuration>
    <property>
        <name>yarn.nodemanager.resource.memory-mb</name>
        <value>2048</value>
    </property>
    <property>
        <name>yarn.nodemanager.resource.cpu-vcores</name>
        <value>3</value>
    </property>
    <property>
        <name>yarn.scheduler.minimum-allocation-mb</name>
        <value>256</value>
    </property>
    <property>
        <name>yarn.scheduler.maximum-allocation-mb</name>
        <value>1024</value>
    </property>
    <property>
        <name>yarn.scheduler.increment-allocation-mb</name>
        <value>256</value>
    </property>
    <property>
      <description>Classpath for typical applications.</description>
      <name>yarn.application.classpath</name>
      <value>
         $HADOOP_CONF_DIR,
         $HADOOP_COMMON_HOME/*,$HADOOP_COMMON_HOME/lib/*,
         $HADOOP_HDFS_HOME/*,$HADOOP_HDFS_HOME/lib/*,
         $HADOOP_MAPRED_HOME/*,$HADOOP_MAPRED_HOME/lib/*,
         $HADOOP_YARN_HOME/*,$HADOOP_YARN_HOME/lib/*
```

```
        </value>
    </property>
    <property>
        <name>yarn.nodemanager.aux-servimces</name>
        <value>mapreduce_shuffle</value>
    </property>
    <property>
        <name>yarn.nodemanager.env-whitelist</name>
        <value>JAVA_HOME,HADOOP_COMMON_HOME,HADOOP_HDFS_HOME, HADOOP_
CONF_DIR,CLASSPATH_PREPEND_DISTCACHE,HADOOP_YARN_HOME,HADOOP_MAPRED_
HOME</value>
    </property>
    <property>
        <name>yarn.resourcemanager.hostname</name>
        <value>master</value>
    </property>
</configuration>
```

6. 从节点列表的配置（主节点）

打开从节点列表配置文件/etc/hadoop/conf.my_cluster/workers，加入从节点的主机名。

```
vim /etc/hadoop/conf.my_cluster/workers
```

该文件看起来如下所示。

```
slave1
slave2
```

7. 复制配置文件到从节点

在主节点上使用 scp 命令从主节点复制配置文件路径到从节点的相同路径。

```
scp -r /etc/hadoop/conf.my_cluster/   slave1:/etc/hadoop/
scp -r /etc/hadoop/conf.my_cluster/   slave2:/etc/hadoop/
```

1.8 Hadoop 的运行

1. 格式化 HDFS（主节点）

在第一次启动 HDFS 之前，使用命令 hdfs namenode –format 格式化 HDFS。

```
sudo -u hdfs hdfs namenode -format
18/09/10 06:47:19 INFO namenode.NameNode: STARTUP_MSG:
/************************************************************************
STARTUP_MSG: Starting NameNode
STARTUP_MSG: user = root
STARTUP_MSG: host = centos7-6.novalocal/192.168.10.21
STARTUP_MSG: args = [-format]
STARTUP_MSG: version = 2.6.0-cdh5.15.1
STARTUP_MSG: classpath = ...
STARTUP_MSG: build=http://github.com/cloudera/hadoop -r 2d822203265a2827554b84cbb46c69b86ccca149;
compiled by 'jenkins' on 2018-08-09T16:22Z
```

```
STARTUP_MSG:   java = 1.8.0_162
************************************************************************/
[...]
/************************************************************************
SHUTDOWN_MSG: Shutting down NameNode at centos7-6.novalocal/ 192.168.10.21
************************************************************************/
```

2. 启动 Hadoop 核心组件

使用 service start 命令启动 Hadoop 核心组件，包括：

- HDFS 命令节点服务：hadoop-hdfs-namenode；
- HDFS 数据节点服务：hadoop-hdfs-datanode；
- YARN 资源管理器服务：hadoop-yarn-resourcemanager；
- YARN 节点管理器服务：hadoop-yarn-nodemanager。

在主节点上启动所有 4 个服务。

```
sudo service hadoop-hdfs-datanode start
sudo service hadoop-hdfs-namenode start
sudo service hadoop-yarn-nodemanager start
sudo service hadoop-yarn-resourcemanager start
starting datanode, logging to /var/log/hadoop-hdfs/hadoop-hdfs- datanode-master.out
Started Hadoop datanode (hadoop-hdfs-datanode):              [  OK  ]

starting namenode, logging to /var/log/hadoop-hdfs/hadoop-hdfs- namenode-master.out
Started Hadoop namenode:                                     [  OK  ]

starting nodemanager, logging to /var/log/hadoop-yarn/yarn-yarn- nodemanager-master.out
Started Hadoop nodemanager:                                  [  OK  ]

starting resourcemanager, logging to /var/log/hadoop-yarn/yarn- yarn-resourcemanager-master.out
Started Hadoop resourcemanager:                              [  OK  ]
```

在从节点上仅启动 HDFS 数据节点和 YARN 节点管理器服务。

```
sudo service hadoop-hdfs-datanode start
sudo service hadoop-yarn-nodemanager start
starting datanode, logging to /var/log/hadoop-hdfs/hadoop-hdfs- datanode-centos7-6.novalocal.out
Started Hadoop datanode (hadoop-hdfs-datanode):              [  OK  ]

starting nodemanager, logging to /var/log/hadoop-yarn/yarn-yarn- nodemanager-centos7-6.novalocal.out
Started Hadoop nodemanager:                                  [  OK  ]
```

3. 建立/tmp 文件夹（主节点）

Hadoop 中的许多组件在运行时会在 HDFS 上生成许多临时文件存储在/tmp 文件夹中。

以 hdfs 用户使用 hdfs dfs -mkdir 命令创建文件夹/tmp 以及 hdfs dfs -chmod 命令将文件夹权限改为任何用户都可以读/写。

```
sudo -u hdfs hdfs dfs -mkdir /tmp
sudo -u hdfs hdfs dfs -chmod -R 1777 /tmp
```

4. 创建用户个人文件夹（主节点）

HDFS 中用户个人文件夹为/user/<用户名>。

以 hdfs 用户使用 hdfs dfs -mkdir 命令创建文件夹/user/root 以及 hdfs dfs -chown 命令将文件夹所有权移交给 root 用户。

```
sudo -u hdfs hdfs dfs -mkdir -p /user/root
sudo -u hdfs hdfs dfs -chown -R root /user/root
```

5. 验证 Hadoop 正常运行（主节点）

通过网页接口的方式验证 Hadoop 状态。打开浏览器，输入 http://master:9870，访问命名节点的网页接口，如图 1-4 所示。

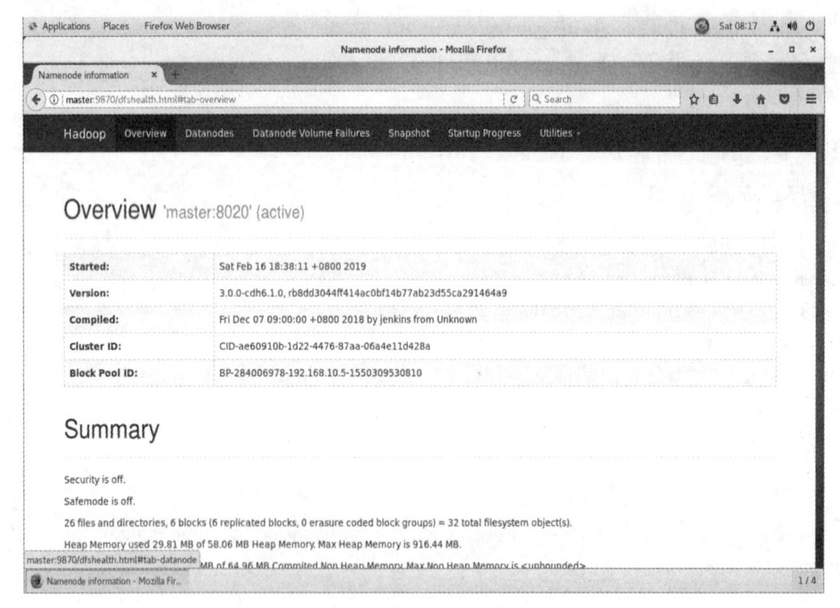

图 1-4　HDFS 命名节点网页界面

切换到最上方的 Datanodes 标签，查看数据节点数量，验证 3 个数据节点是否正常运行，如图 1-5 所示。

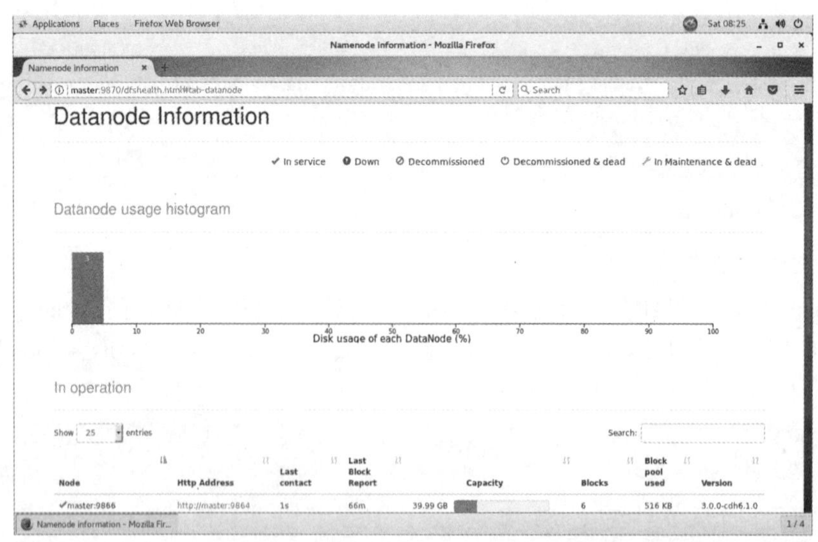

图 1-5　HDFS 命名节点网页界面

如果在无图形界面版操作系统上没有浏览器，则以 hdfs 用户使用 hdfs dfsadmin -report 命令查看 3 个数据节点是否正常运行。

```
sudo -u hdfs hdfs dfsadmin -report
Configured Capacity: 5906622726144 (5.37 TB)
Present Capacity: 5462765572096 (4.97 TB)
DFS Remaining: 5462765494272 (4.97 TB)
DFS Used: 77824 (76 KB)
DFS Used%: 0.00%
Replicated Blocks:
        Under replicated blocks: 0
        Blocks with corrupt replicas: 0
        Missing blocks: 0
        Missing blocks (with replication factor 1): 0
        Low redundancy blocks with highest priority to recover: 0
        Pending deletion blocks: 0
Erasure Coded Block Groups:
        Low redundancy block groups: 0
        Block groups with corrupt internal blocks: 0
        Missing block groups: 0
        Low redundancy blocks with highest priority to recover: 0
        Pending deletion blocks: 0

-------------------------------------------------
Live datanodes (3):

Name: 192.168.95.128:9866 (master)
Hostname: master
Decommission Status : Normal
Configured Capacity: 1968874242048 (1.79 TB)
DFS Used: 28672 (28 KB)
Non DFS Used: 47915716608 (44.62 GB)
DFS Remaining: 1820921831424 (1.66 TB)
DFS Used%: 0.00%
DFS Remaining%: 92.49%
Configured Cache Capacity: 0 (0 B)
Cache Used: 0 (0 B)
Cache Remaining: 0 (0 B)
Cache Used%: 100.00%
Cache Remaining%: 0.00%
Xceivers: 1
Last contact: Mon Feb 25 06:47:31 UTC 2019
Last Block Report: Mon Feb 25 06:00:05 UTC 2019

Name: 192.168.95.129:9866 (slave1)
Hostname: slave1
Decommission Status : Normal
Configured Capacity: 1968874242048 (1.79 TB)
DFS Used: 24576 (24 KB)
Non DFS Used: 47915720704 (44.62 GB)
DFS Remaining: 1820921831424 (1.66 TB)
DFS Used%: 0.00%
```

```
DFS Remaining%: 92.49%
Configured Cache Capacity: 0 (0 B)
Cache Used: 0 (0 B)
Cache Remaining: 0 (0 B)
Cache Used%: 100.00%
Cache Remaining%: 0.00%
Xceivers: 1
Last contact: Mon Feb 25 06:47:31 UTC 2019
Last Block Report: Mon Feb 25 06:41:54 UTC 2019

Name: 192.168.95.130:9866 (slave2)
Hostname: slave2
Decommission Status : Normal
Configured Capacity: 1968874242048 (1.79 TB)
DFS Used: 24576 (24 KB)
Non DFS Used: 47915720704 (44.62 GB)
DFS Remaining: 1820921831424 (1.66 TB)
DFS Used%: 0.00%
DFS Remaining%: 92.49%
Configured Cache Capacity: 0 (0 B)
Cache Used: 0 (0 B)
Cache Remaining: 0 (0 B)
Cache Used%: 100.00%
Cache Remaining%: 0.00%
Xceivers: 1
Last contact: Mon Feb 25 06:47:31 UTC 2019
Last Block Report: Mon Feb 25 06:44:02 UTC 2019
```

打开浏览器，输入 http://master:8088，访问资源管理器的网页接口，验证 3 个节点管理器是否正常运行，如图 1-6 所示。

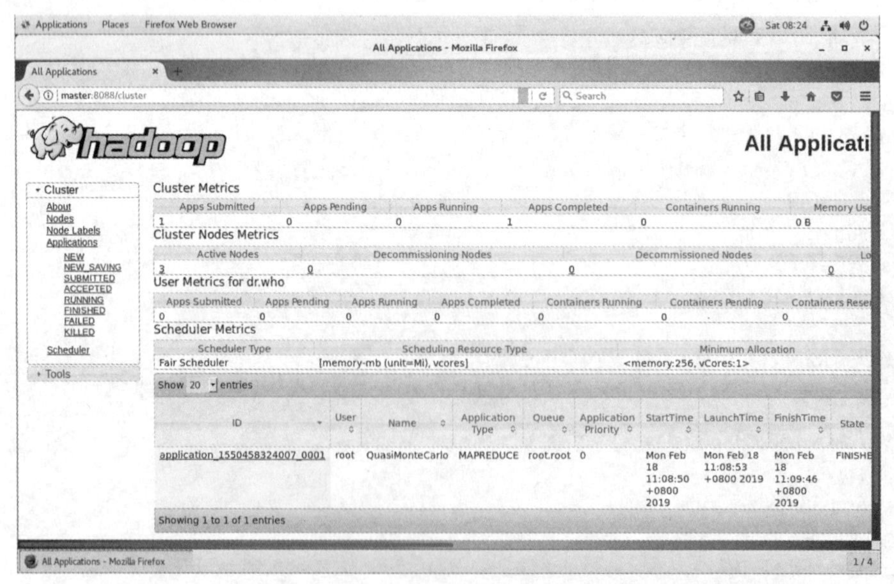

图 1-6　YARN 资源管理器网页界面

如果在无图形界面版操作系统上没有浏览器，则使用 yarn node -list 命令查看 3 个节点管理器是否正常运行。

```
yarn node -list
2019-02-25 06:53:42,039 INFO client.RMProxy: Connecting to ResourceManager at master/172.25.0.2:8032
Total Nodes:3
     Node-Id          Node-State     Node-Http-Address        Number-of-Running-Containers
     slave2:34366     RUNNING        slave2:8042              0
     slave1:35647     RUNNING        slave1:8042              0
     master:43798     RUNNING        master:8042              0
```

提交一个计算π的示例 MapReduce 程序，第 1 个参数表示 Map 任务数，第 2 个参数表示每个 Map 任务的样本数。这里不需要关心具体语法，在之后的章节中会详细介绍。

```
hadoop jar /usr/lib/hadoop-mapreduce/hadoop-mapreduce-examples.jar pi 2 100
Number of Maps    = 2
Samples per Map = 100
Wrote input for Map #0
Wrote input for Map #1
Starting Job
[...]
Job Finished in 29.48 seconds
Estimated value of Pi is 3.12000000000000000000
```

小　结

本章首先介绍了 Hadoop 的产生背景、简要历史、生态系统组件、版本和商用支持。Hadoop 是一个大数据存储、处理和挖掘的软件生态系统，能够采用分布式的方式处理 PB 级别的数据。狭义的 Hadoop 包含 Hadoop Common、HDFS、YARN 和 MapReduce 4 个组件，而广义的 Hadoop 生态系统包含 Hive、Spark、Sqoop、Oozie 等软件工具。

本章还着重介绍了分布式 Hadoop 的安装流程，包括基础环境的配置（如防火墙、主机名映射、时钟同步和免密登录）、软件包的安装（命名节点、数据节点、资源管理器、节点管理器和关系型数据库）、软件包的配置和验证安装结果。

习　题

1. Hadoop 与传统数据处理软件有什么区别？
2. Hadoop 生态系统中的常用组件有哪些？主要功能有哪些？
3. Hadoop 的商用公司都有哪些？业务模式如何？
4. 熟悉分布式 Hadoop 的安装流程，搭建 Hadoop 环境。
5. 验证 Hadoop 安装是否成功，并运行测试程序。

第 2 章

HDFS 文件系统 <<<

2.1 HDFS 简介

Hadoop 分布式文件系统（Hadoop Distributed File System，HDFS）具有高容错性，并能部署在商用服务器上，提供高吞吐量的数据访问，十分适用于大数据应用。HDFS 的设计初衷和假设主要有以下几点。

- 高容错性：硬件故障是常态而不是异常。HDFS 集群可以由成百上千台物理服务器组成，每台都存储了文件系统的一部分数据。这意味着 HDFS 中的某一部分总有可能出现故障。因此，HDFS 的设计目标就包括及时检测故障并从中恢复。
- 高吞吐、高延时：HDFS 更多用于数据批处理而不是交互式处理，即强调数据访问的高吞吐量，但响应速度较低。
- 大数据集：HDFS 存储的通常为大数据集，从几 GB 到几 TB 不等，这样才能充分体现 HDFS 的优势。一个 HDFS 集群可以支持千万量级的文件数量。
- “写一次读多次”的数据模型：一个文件一旦创建，只能在文件末尾追加或截断操作，但是不支持在文件任意位置做修改。这一简单的数据模型简化了数据一致性的问题，并且提升了数据处理的吞吐量。
- 计算本地化：使得数据处理发生在靠近数据存储的节点能够减小网络开销并提高系统吞吐量，而不是将数据迁移到某个应用再进行处理。

2.2 HDFS 架构

HDFS 采用主从架构，由一个命名节点（NameNode）和多个数据节点（DataNode）组成。命名节点用于管理文件系统的命名空间和文件的访问权限等。数据节点用于管理它们所运行的节点的数据存储，通常每个节点运行一个数据节点进程。HDFS 暴露给用户的接口只是一个文件的命名空间。就内部机制而言，一个文件会分割成多个文件块（block），每个文件块又会存储在多个数据节点。命名节点仅仅执行文件系统的命名空间操作，如打开、关闭和重命名文件和文件夹，并且决定文件块与数据节点的映射关系。而数据节点会接受命名节点的指令，真正响应客户端的文件读写请求，做文件块的创建、删除和复制。HDFS 架构如图 2-1 所示。

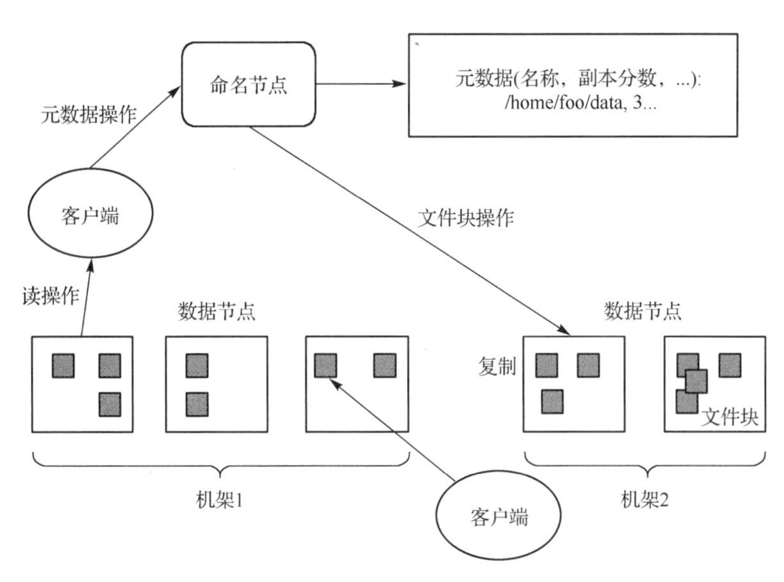

图 2-1 HDFS 架构图

HDFS 与传统文件系统的组织方式类似，都是层级式的，即可以在文件夹中创建子文件夹或存储文件。用户可以创建、删除、移动和重命名文件，但是暂时不支持软连接和硬连接。HDFS 中可以限制用户占用空间和访问权限。

2.3 HDFS 文件块

HDFS 能够可靠地在集群中存储大数据集，每个文件存储成多个文件块，每个文件块复制多份以达到高容错性。文件块的大小和副本份数都可以配置。命名节点决定所有文件副本的映射。数据节点会周期性地发送心跳和文件块报告给命名节点，心跳表明数据节点运行正常，文件块报告包含数据节点中的所有文件块列表。

如图 2-2 所示，文件 part-0 设置为副本 2 份，有 2 个文件块，分别为 b_1 和 b_3，其中：

- 文件块 b_1 分别存储在节点 n_1 和 n_3 上；
- 文件块 b_3 分别存储在节点 n_5 和 n_7 上。

文件 part-1 设置为副本 3 份，有 3 个文件块，分别为 b_2、b_4 和 b_5，其中：

- 文件块 b_2 分别存储在节点 n_1、n_2 和 n_4 上；
- 文件块 b_4 分别存储在节点 n_3、n_6 和 n_8 上；
- 文件块 b_5 分别存储在节点 n_4、n_5 和 n_7 上。

数据副本的放置位置对于 HDFS 的可靠性和性能至关重要。HDFS 采用了一种叫做机架感知的策略。数据中心通常放置了多个机架（又称机柜），每个机架中又放置了多台物理服务器，同一个机架中的服务器共享同一个交换机。通常情况下，同一个机架内的网络带宽远大于不同机架之间的网络带宽，意味着数据在同一个机架的不同服务器之间的传输速度要远远大于在不同机架的不同服务器之间的传输速度。

命名节点能够检测每个数据节点所在的机架。一种简单但并非最佳的策略，是将数据副本放置在不同的机架上，这样可以防止整个机架故障导致的数据丢失，并能在读取数据时充分利用多个机架的网络带宽。对于副本份数为 3 的情况，HDFS 采取的策略是，将 2 份副本放置在同一个机架的不同数据节点，第 3 份副本放置在另一个机架的数据节点。

命名节点

文件名	副本分数	文件块标识符
part-0	2	b_1, b_3
part-1	3	b_2, b_4, b_5

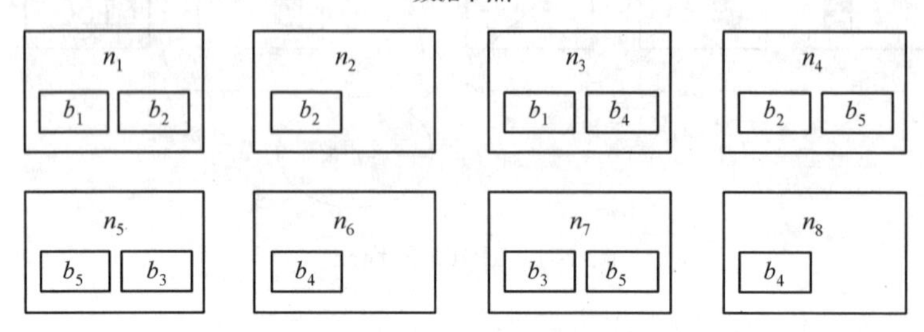

图 2-2　HDFS 文件块

命名节点不会将 2 份副本放置在同一个数据节点上，因此最大副本份数为集群中数据节点的数量。

HDFS 文件块的大小和副本份数可以在配置文件 hdfs-size.xml 中设置：

- dfs.blocksize 表示文件块的大小，默认为 128m，即 128 MB；
- dfs.replication 表示文件块的副本份数，默认为 3。

2.4　HDFS 常用操作

HDFS 提供类似 Shell 的命令操作文件，形如 hdfs dfs -<子命令>，具体语法可参考 https://hadoop.apache.org/docs/stable/hadoop-project-dist/hadoop-common/FileSystemShell.html 文档。

HDFS 中也可以使用相对路径，当前工作目录为 HDFS 上的家目录，即/user/<用户名>。HDFS 中常用子命令如下所示：

- -appendToFile：追加写入文件；
- -cat：显示文件内容；
- -cp：在 HDFS 中复制文件（夹）；
- -get：从 HDFS 复制文件（夹）到本地文件系统；
- -ls：显示文件（夹）统计信息；
- -mv：在 HDFS 中移动文件（夹）；
- -mkdir：创建文件夹；
- -put：从本地文件系统复制文件（夹）到 HDFS；
- -rm：删除文件（夹）。

使用 hdfs dfs -mkdir 命令在 HDFS 上创建文件夹 test，这里使用了相对路径，其绝对路径为/user/<用户名>/test。

```
hdfs dfs -mkdir test
```

使用 hdfs dfs -put 命令从本地文件系统复制文件（夹）到 HDFS，其中：

- 第 1 个参数表示本地源文件，文件名可以使用通配符；
- 第 2 个参数表示 HDFS 目标文件夹。

这里将 data/novels 本地文件夹中的所有文件复制到 HDFS 的 test 文件夹中。

```
hdfs dfs -put data/novels/* test
```

使用 hdfs dfs -ls 命令显示文件（夹）统计信息。

- 对于文件，显示的统计信息格式为：

```
<权限> <副本份数> <用户名> <用户组名> <文件大小> <修改日期> <修改时间> <文件名>
```

- 对于文件夹，显示的统计信息格式为：

```
<权限> <用户名> <用户组名> <修改日期> <修改时间> <文件夹名>
```

例如：

```
hdfs dfs -ls test
Found 3 items
-rw-r--r--    1 root hadoop      389386 2018-09-28 11:52 test/Woolf_Lighthouse_1927.txt
-rw-r--r--    1 root hadoop      952293 2018-09-28 11:52 test/Woolf_Night_1919.txt
-rw-r--r--    1 root hadoop      730235 2018-09-28 11:52 test/Woolf_Years_1937.txt
```

使用 hdfs dfs -cat 命令显示文件内容。由于内容较多，在命令末尾添加管道和 head 命令仅显示前几行。

```
hdfs dfs -cat test/Woolf_Years_1937.txt | head
The Years

1880

It was an uncertain spring. The weather, perpetually changing,
sent clouds of blue and of purple flying over the land. In the
country farmers, looking at the fields, were apprehensive; in
London umbrellas were opened and then shut by people looking up at
the sky. But in April such weather was to be expected. Thousands
```

使用 hdfs dfs -cp 命令在 HDFS 中复制文件（夹），其中：

- 第 1 个参数表示 HDFS 源文件（夹）；
- 第 2 个参数表示 HDFS 目标文件（夹）。

```
hdfs dfs -cp test/Woolf_Years_1937.txt test/Woolf_Years_1937_copy.txt
```

使用 hdfs dfs -get 命令从 HDFS 复制文件（夹）到本地文件系统。

```
hdfs dfs -get test/Woolf_Years_1937_copy.txt
```

使用 hdfs dfs -rm 命令删除文件（夹），选项-r 表示删除文件夹和文件夹下的所有文件。

```
hdfs dfs -rm test/Woolf_Years_1937_copy.txt
hdfs dfs -rm -r test
```

小　　结

本章主要介绍了 Hadoop 分布式文件系统 HDFS。HDFS 的架构主要包含了命名节点（主节点）和数据节点（从节点）。命名节点用于管理文件系统的命名空间和文件的访问权限等。数据节点用于管理它们所运行的节点的数据存储，通常每个节点运行一个数据节点进程。一个文件会分割成多个文件块，每个文件块又会存储在多个数据节点中。

本章还着重介绍了 HDFS 文件操作，包括文件的显示、复制、移动、上传、下载等命令形式与 Linux Shell 十分相似。

习　　题

1. HDFS 架构的特点有哪些？
2. HDFS 的文件块是什么？如何设置文件块大小？
3. 在 HDFS 中执行文件（夹）的显示、复制、移动、上传、下载等操作。

第3章

YARN 资源管理 ‹‹‹

3.1 YARN 架构

YARN 的核心功能是集群资源管理以及应用的调度监控，应用可以是任何类型的，如 MapReduce、Spark 或 Tez 程序。用户大多时候不需要直接与 YARN 进行交互，而只需要调用应用的接口完成特定任务，例如用户直接提交 MapReduce 程序，而 MapReduce 计算框架会进一步与 YARN 做交互，但这些资源管理细节已经对用户透明。

首先需要理解应用（application）、作业（job）和任务（task）的关系。Hadoop 集群中运行着用户的应用，应用可以是一个作业，也可以是一系列作业。由于 Hadoop 采用主从架构，一个作业包含了若干并行化的任务，分配到不同节点执行。例如，一个在 Spark 中执行的 k 均值聚类算法是一个应用，由于该算法需要遍历数据多次，每次迭代都是一个作业，而每个作业又包含多个并行化的任务在不同节点执行。

与 HDFS 类似，YARN 同样采用主从架构，由一个资源管理器（ResourceManager）和多个节点管理器（NodeManager）组成。资源管理器用于裁决集群中所有应用能够使用的资源。节点管理器用于监控它们所运行的节点中容器（container）的资源使用情况，包括 CPU、内存、磁盘和网络带宽，并上报给资源管理器。

每个应用都有一个应用主程序（ApplicationMaster），用于与资源管理器协商资源，并与节点管理器协作执行和监控具体任务。具体的任务都在容器中执行，每个容器对应一定的资源（CPU 和内存）。

应用提交和执行的流程如图 3-1 所示。主节点 n_0 运行资源管理器，从节点 n_1、n_2 和 n_3 运行节点管理器，节点管理器会定期向资源管理器上报节点的资源状态。首先，客户端向资源管理器提交应用，资源管理器会选择一个从节点为启动一个应用主程序。应用主程序根据应用运行状况，可以动态地向资源管理器申请资源，资源以容器的形式进行分配，任务最终会在多个容器中并行执行，容器会定期向应用主程序上报任务执行状态。

资源管理器主要由调度器和应用管理器两部分组成。

- 调度器（Scheduler）：用于根据排队策略和资源的限制，将资源分配给集群上运行的多个应用。这里的调度器只负责作业调度，并不负责监控应用执行状态，也不会重启因硬件故障或应用失败所导致的失败任务。调度器调度作业时，以容器为基本单位，每个容器可以对应不同的 CPU、内存、磁盘和网络带宽。

- 应用管理器（ApplicationManager）：用于接受作业提交，协商第一个容器用作应用主程序，并在发生故障时重启应用主程序。与具体应用相关的管理工作都交给应用主程序，如与调度器协商容器资源、监控状态和进度。

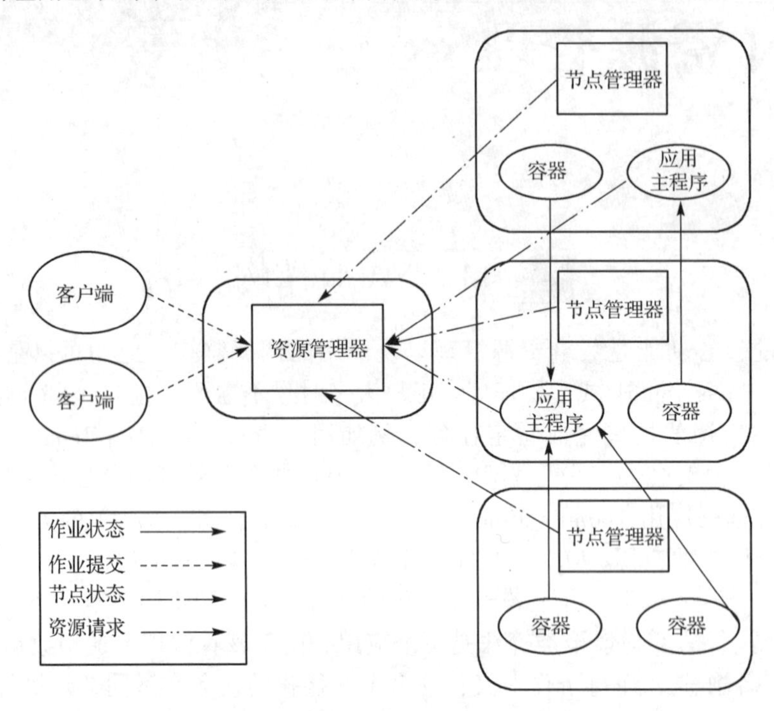

图 3-1　应用提交和执行的流程图

3.2　YARN 调度策略

理想情况下，应用会立即取得它们请求的任何资源。然而现实中，对于一个繁忙的集群而言，资源是有限的，应用通常都要等待一段时间才能取得请求的资源。YARN 调度器会根据调度策略将资源分配给应用。没有哪种调度策略是最佳的，因此 YARN 提供了多种调度器和可配置的策略供选择。

YARN 提供 3 种调度器：先进先出（FIFO）调度器、性能（Capacity）调度器和公平（Fair）调度器。先进先出调度器将提交的应用放在一个队列中，按照应用提交的顺序执行。最先提交的应用也最先取得资源，等到它的资源请求完全被满足，队列中的下一个应用开始执行。

先进先出调度器的优点是，机制简单容易理解，并且不需要做任何其他配置。但缺点也很明显，不适合用于多用户共享的集群，如大型企业的实际应用场景。资源消耗极大的应用会长时间占用集群的所有资源，使得一些资源消耗极小的应用也不得不长时间等待。在多用户共享的集群中，大多采用性能调度器或公平调度器。这两种调度策略使得资源消耗较大的应用也能及时完成，同时使得用户能够并行执行一些资源消耗较小的交互式查询，及时返回结果。

图 3-2～图 3-4 展示了 3 种调度器的区别。应用 1 资源消耗较大，应用 2 资源消耗较小。

① 在先进先出调度器中，应用 2 不得不等待应用 1 完成才能开始执行。

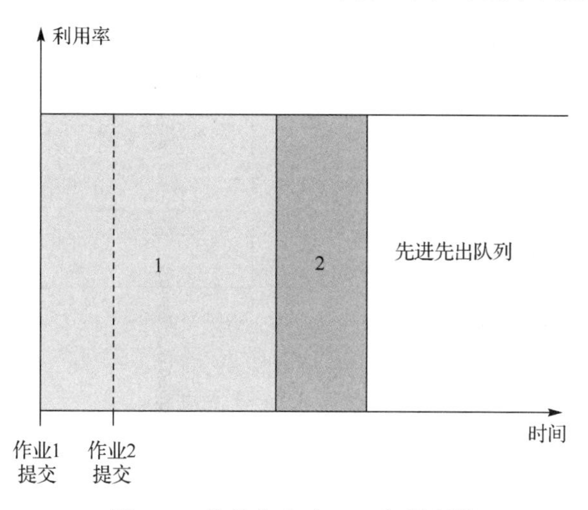

图 3-2　先进先出（FIFO）调度器

② 在性能调度器中，配置了 2 个队列，队列 *A* 用于执行资源消耗大的应用，队列 *B* 用于执行资源消耗小的应用。应用 2 一提交则开始执行，但相应的代价是，即使在队列 *B* 空闲的情况下，应用 1 也无法利用队列 *A* 中的资源，对集群的资源利用率产生一定影响。

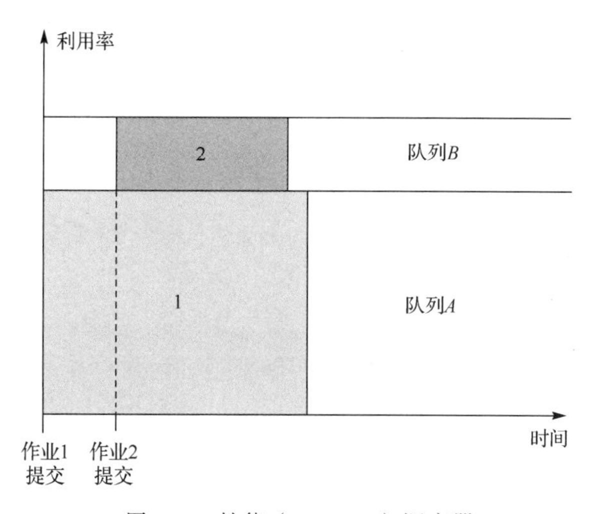

图 3-3　性能（Capacity）调度器

③ 在公平调度器中，并不预留任何资源，而是动态调整资源分配。当集群中只有应用 1 在执行时，它可以占用所有资源。当应用 2 提交后，一半资源分配给了应用 2。当应用 2 执行结束后，应用 1 又可以占用所有资源。需要注意的是，应用 2 并不一定马上能够得到集群的一半资源，而必须等到应用 1 释放部分容器，因此会有一定的延时。

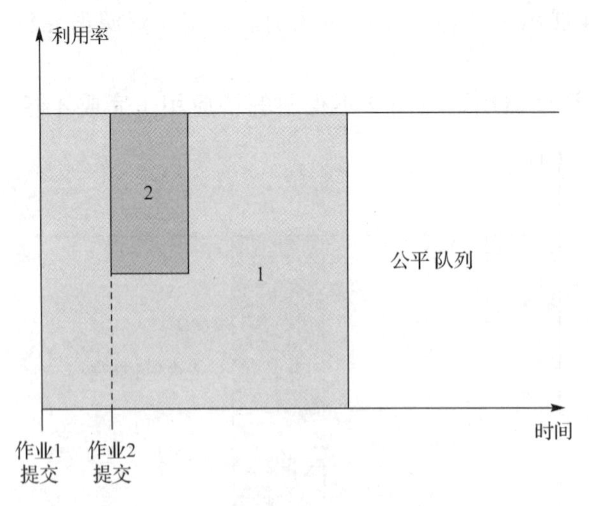

图 3-4　公平（Fair）调度器

3.3　YARN 常用操作

YARN 提供一系列命令用于操作应用，具体语法可以参考 https://hadoop.apache.org/docs/stable/hadoop-yarn/hadoop-yarn-site/YarnCommands.html 文档。

1. 提交应用

YARN 是一个通用的资源调度软件，可以有多种方式提交应用。yarn jar <JAR 文件> [<主类名>] <参数> ...命令用于提交 JAR 文件类型的应用。

以下例子提交一个计算π的示例 MapReduce 程序，其中：

- JAR 文件路径为/usr/lib/hadoop-mapreduce/hadoop-mapreduce-examples.jar；
- 主类名为 pi；
- 第 1 个参数表示 Map 任务数为 2，第 2 个参数表示每个 Map 任务的样本数为 100。

```
yarn jar /usr/lib/hadoop-mapreduce/hadoop-mapreduce-examples.jar pi 2 100
Number of Maps   = 2
Samples per Map = 100
Wrote input for Map #0
Wrote input for Map #1
Starting Job
[...]
Job Finished in 29.48 seconds
Estimated value of Pi is 3.12000000000000000000
```

2. 查看或杀死应用

yarn application -<子命令>命令用于查看或杀死应用。常用子命令如下所示：

- -list：列出资源管理器中的应用，可以进一步使用 appTypes 或 appStates 子命令进行筛选；
- appStates：按照应用状态做筛选，可以传入多个以逗号分隔的状态，常用的状态有 ALL（所有）、SUBMITTED（已提交）、ACCEPTED（已接收）、RUNNING（运行中）、FINISHED（已完成）、FAILED（已失败）和 KILLED（已杀死）；

- appTypes：按照应用类型做筛选，可以传入多个以逗号分隔的类型；
- -kill：杀死应用标识符 ApplicationId 所对应的应用；
- -status：查看应用标识符 ApplicationId 所对应的应用状态。

使用 pyspark 命令启动一个 Spark 应用。

```
pyspark --master yarn
```

新启动一个终端，使用 yarn application -list 命令列出所有应用，默认列出状态为 SUBMITTED（已提交）、ACCEPTED（已接收）和 RUNNING（运行中）的应用。

```
yarn application -list
18/10/01 10:01:05 INFO client.RMProxy: Connecting to ResourceManager at /0.0.0.0:8032
Total number of applications (application-types: [] and states: [SUBMITTED, ACCEPTED, RUNNING]):1
Application-Id          Application-Name        Application-Type    User        Queue
    State               Final-State             Progress            Tracking-URL
application_1538355655973_0002                  PySparkShell        SPARK       root root.root
    RUNNING             UNDEFINED               10%                 http://192.168.10.40:4040
```

可以看出，该应用的应用标识符为 application_1538355655973_0002。

使用选项 appStates 进一步选择列出已完成（FINISHED）的应用。

```
yarn application -list -appStates FINISHED
18/10/01 10:28:48 INFO client.RMProxy: Connecting to ResourceManager at /0.0.0.0:8032
Total number of applications (application-types: [] and states: [FINISHED]):1
Application-Id                  Application-Name        Application-Type    User Queue
    State       Final-State     Progress            Tracking-URL
application_1538355655973_0001  QuasiMonteCarlo         MAPREDUCE           root root.root
    FINISHED SUCCEEDED          100%    http://123.novalocal:19888/jobhistory/job/job_1538355655973_0001
```

使用 yarn application -status 查看应用状态，后跟之前得到的正在运行的应用标识符。

```
yarn application -status application_1538355655973_0002
Application Report :
    Application-Id : application_1538355655973_0002
    Application-Name : PySparkShell
    Application-Type : SPARK
    User : root
    Queue : root.root
    Start-Time : 1538387957263
    Finish-Time : 0
    Progress : 10%
    State : RUNNING
    Final-State : UNDEFINED
    Tracking-URL : http://192.168.10.40:4040
    RPC Port : 0
    AM Host : 192.168.10.40
    Aggregate Resource Allocation : 4212095 MB-seconds, 5481 vcore-seconds
    Log Aggregation Status : DISABLED
    Diagnostics :
```

使用 yarn application-kill 杀死应用，后跟之前得到的正在运行的应用标识符。

```
yarn application -kill application_1538355655973_0002
18/10/01 10:59:10 INFO client.RMProxy: Connecting to ResourceManager at /0.0.0.0:8032
Killing application application_1538355655973_0002
18/10/01 10:59:12 INFO impl.YarnClientImpl: Killed application application_1538355655973_0002
```

切换到启动 Spark 的终端，可以看到如下消息表示该应用已经被杀死。

```
18/10/01 10:59:12 ERROR cluster.YarnClientSchedulerBackend: Yarn application has already exited with state KILLED!
```

3. YARN 网页界面

打开浏览器，输入 http://localhost:8088，访问资源管理器网页界面，如图 3-5 所示。

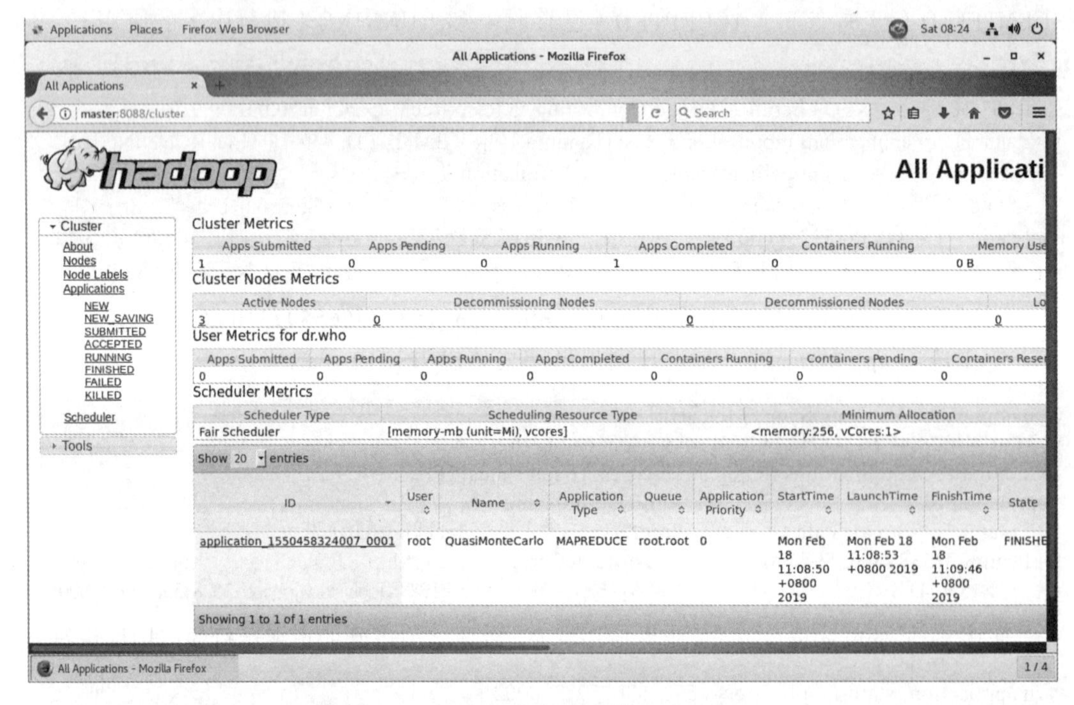

图 3-5　YARN 资源管理器网页界面

在最左侧菜单，可以分别查看：

- 资源管理器信息（About）；
- 集群节点（Nodes）；
- 应用（Applications）；
- 调度器（Scheduler）。

小　结

本章主要介绍了 Hadoop 集群资源管理工具 YARN。YARN 的架构主要包含了资源管理器（主节点）和节点管理器（从节点）。资源管理器用于裁决集群中所有应用能够使用的资源。节点管理器用于监控它们所运行的节点中容器的资源使用情况，包括 CPU、内存、磁盘和网络带宽，并上报给资源管理器。

本章还着重介绍了 YARN 常用操作，包括应用的提交、查看和终止。YARN 调度策略包括先进先出调度器、性能调度器和公平调度器。

习　题

1. YARN 架构的特点有哪些？
2. YARN 的调度策略有哪些？各有什么特点？
3. 描述应用提交到 YARN 上的整个流程。
4. 描述应用、作业和任务的关系。
5. 在 YARN 中执行应用的提交、查看和终止等操作。

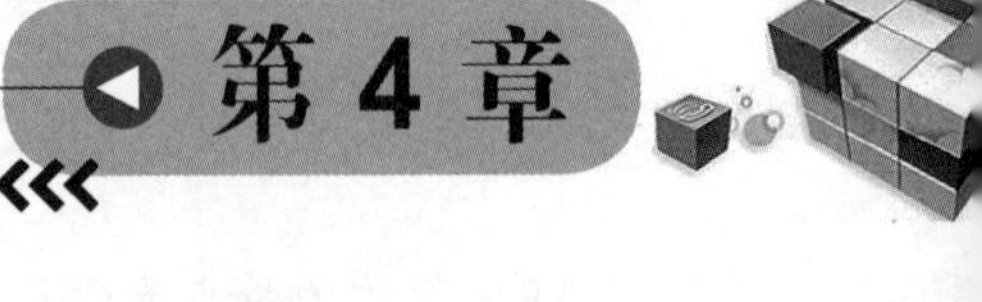

MapReduce 计算框架

4.1 MapReduce 原理

MapReduce 是一个软件编程框架，用于开发能够在商用硬件组成的集群中并行处理大数据的应用程序。为了理解方便，本节以词频统计为例，详细解读 MapReduce 原理。整个流程如图 4-1 所示。

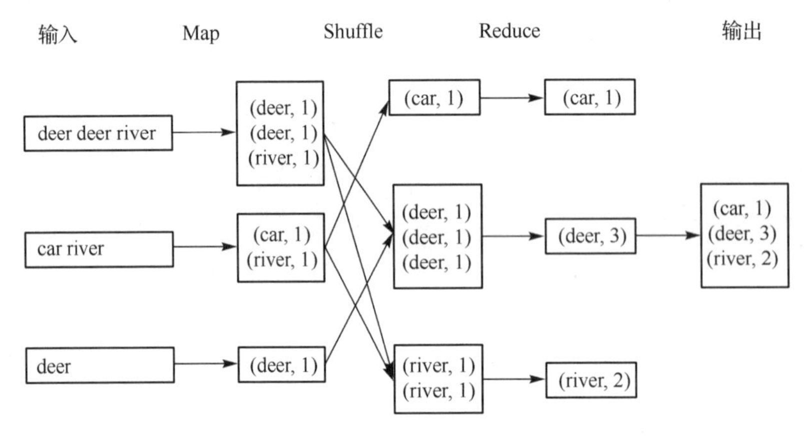

图 4-1 MapReduce 原理图

MapReduce 的基本原理是，将数据处理分为 2 个阶段，即 Map 阶段和 Reduce 阶段。每一阶段的输入和输出都是一个键值对，具体数据类型可以由用户指定。用户还需要实现 2 个函数，即 map( )和 reduce( )函数。

Map 阶段的输入为 3 行文本，输入格式即为文本格式。文件的每一行都会调用一次 map( )函数，输入的键为自文件开头到当前行的偏移字节数，输入的值为每一行的文本。在本例中，map( )函数很简单，只需要将每一行文本按空格分隔成单词，输出键为单词、值为 1 的键值对。Reduce( )函数统计每个单词出现的次数。

具体而言，输入文本为：

```
deer deer river
car river
deer
```

map( )函数的输入键值对为：

```
(0, deer car river)
(14, deer river)
(24, deer)
```

Map( )函数将输入文本按空格分隔成单词作为键、1 为值，输出的键值对为：

```
(deer, 1)
(car, 1)
(river, 1)
(deer, 1)
(river, 1)
(deer, 1)
```

map( )函数的输出会经过 Shuffle 操作，即按键排序和分组，将结果输入 reduce( )函数。reduce( )函数的输入键值对为：

```
(car, [1])
(deer, [1, 1, 1])
(river, [1, 1])
```

即每个单词都对应了一个由 1 组成的列表，每个 1 代表该单词出现过一次。而在后续章节介绍的 Hadoop 流处理中，相同键所对应的值并不一定会组成所用语言中的列表数据结构，但还是会按键排序，即：

```
(car, 1)
(deer, 1)
(deer, 1)
(deer, 1)
(river, 1)
(river, 1)
```

reduce( )函数只需要遍历列表，或在 Hadoop 流处理中判断当前键是否与上一个键一致，对相同键对应的值求和，输出的键值对为：

```
(car, 1)
(deer, 3)
(river, 2)
```

后续章节会给出本例 MapReduce 程序的具体实现。在此之前，还需要理解 MapReduce 作业数据流。

通常情况下，MapReduce 作业的计算节点和数据的存储节点是同一系列节点。换言之，Hadoop 中的每个从节点一般既作为 YARN 资源池的一部分，提供 CPU、内存等资源，同时也作为 HDFS 数据节点的一部分，提供存储。MapReduce 作业在向 YARN 申请资源时，会尽可能在该任务输入数据所在节点上执行，以节省网络带宽。

4.2 MapReduce 作业数据流

一个 MapReduce 作业由输入数据、MapReduce 程序和配置信息组成。Hadoop 在运行 MapReduce 作业时，会分成多个 Map 和 Reduce 任务。YARN 负责调度这些任务，并在集群中的节点上执行，任务如果失败则会在不同的节点上重试。

Hadoop 将 MapReduce 作业的输入分隔成多个固定大小的数据划分（split）。对于每个数据划分，都启动一个 Map 任务，用于在输入数据中的每条记录上调用 map( ) 函数。多数情况下，数据划分的大小与文件块的大小一致，默认为 128 MB。

Hadoop 会尽其所能将 Map 任务在该任务输入数据所在节点上执行，以节省网络带宽，称为“数据本地化优化”。如果该任务输入数据副本的所有节点都没有空闲资源，调度器则会尽量将任务在同一机架的其他节点执行。如果同一机架的所有节点仍然没有空闲资源，才考虑在其他机架的节点执行该任务。

Map 任务的输出直接写入本地磁盘，而不是 HDFS。究其原因，Map 任务的输出只是中间结果，会输出 Reduce 任务做进一步处理进而产生最终结果，整个作业完成时 Map 任务的输出会被删除。假如将此中间结果存储在 HDFS 中形成多份副本，则会造成过多的系统开销。

Reduce 任务不具备 Map 任务的数据本地化优势，一个 Reduce 任务的输入通常情况下来自所有 Map 任务的输出。因此，Map 任务的输出经过按键排序后，必须通过网络传输到执行 Reduce 任务的节点，该过程又称 Shuffle 操作。同一个键的键值对会合并成列表，输入用户定义的 reduce( )函数。Reduce 任务的输出通常情况下会写入 HDFS。

Reduce 任务的数量并不取决于输入数据大小，因为在 MapReduce 作业启动前，框架无法预知 Map 任务的输出数据大小。Reduce 任务的数量通常情况下由用户或更高层的应用指定。当指定了多个 Reduce 任务时，Map 任务会将输出按键做分区（partition），每个分区对应一个 Reduce 任务。每个分区可以对应多个键，但每个键只能对应一个分区，即键相同的记录必然进入同一个 Reduce 任务。默认的分区方法为，对键执行哈希函数，哈希值相同的记录为同一个分区，用户也可以指定自己的分区函数。整个流程如图 4-2 所示。

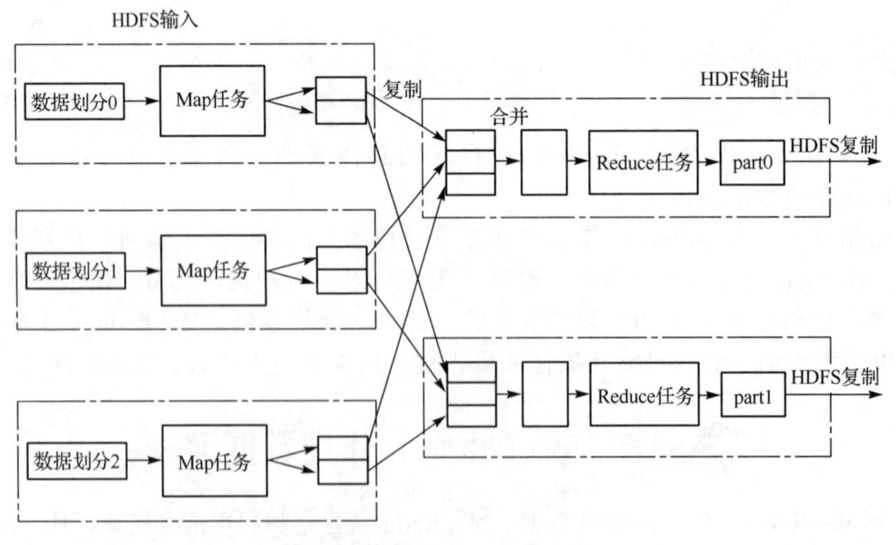

图 4-2　MapReduce 作业流程图

4.3 Hadoop 流处理

Hadoop 的接口函数语言为 Java。Hadoop 流处理（streaming）使得用户能够采用 UNIX 标准流作为接口与 Hadoop 交互，而不必须使用 Java。用户可以通过读取标准输入流和写入标准输出流的方式，使用任何编程语言开发 MapReduce 程序。本章所有 MapReduce 程序将基于 Python 编写，通过 Hadoop 流处理执行。

Hadoop 流处理最基本的调用语法为

```
hadoop jar hadoop-streaming.jar \
  -input <输入路径> \
  -output <输出路径> \
  -mapper <'map'函数脚本> \
  -reducer <'reduce'函数脚本>
```

4.4 MapReduce 程序实现词频统计

1. map( )函数

创建 map( )函数的 Python 脚本为 mapper.py。脚本读取标准输入，遍历每一行文本。调用 strip()函数去除字符串首尾不可见字符，并调用 lower()函数将文本转换为小写，再调用程序包 re 中的 split()函数将每一行文本按非字母的字符分隔成单词。对于每一个分隔得到的单词，如果长度大于 0，输出以单词为键、以 1 为值的键值对，以制表符做分隔符，调用 print()函数写入标准输出。

```
#!/usr/bin/env python
import sys, re
for line in sys.stdin:
    line = line.strip( ).lower( )
    keys = re.split("[^a-z]", line)
    for key in keys:
        if len(key) > 0:
            value = 1
            print( "%s\t%d" % (key, value) )
```

2. reduce( )函数

创建 reduce( )函数的 Python 脚本为 reducer.py。脚本首先定义变量 last_key 存储上一行的键（即单词），running_total 存储单词的出现频次。读取标准输入，遍历每一行文本。调用 strip()函数去除字符串首尾不可见字符，再调用 split()函数将每行文本按制表符分隔成单词和出现频次（本例中都是 1）。由于 Reduce 任务的输入都已经按照键做了排序，如果当前行的键与上一行的键相同，则将词频累加，如果不相同，则表示上一行的键已经全部统计完毕，因此调用 print()函数将键和累加的词频写入标准输出。最后 2 行代码则用于输出最后一个键和对应的词频。

```
#!/usr/bin/env python
import sys
```

```
last_key = None
running_total = 0
for input_line in sys.stdin:
    input_line = input_line.strip( )
    this_key, value = input_line.split("\t", 1)
    value = int(value)
    if last_key == this_key:
        running_total += value
    else:
        if last_key:
            print( "%s\t%d" % (last_key, running_total) )
            running_total = value
            last_key = this_key
        if last_key == this_key:
            print( "%s\t%d" % (last_key, running_total) )
```

3．本地测试

在提交到 Hadoop 执行前，先在本地进行测试。

使用 chmod 命令开放脚本的执行权限。

```
chmod 755 mapper.py
chmod 755 reducer.py
```

根据之前章节的例子，使用 echo 命令和输出重定向，创建一个仅有 3 行内容的文本文件 test.txt。

```
echo "deer deer river
car river
deer" > test.txt
```

使用 cat 命令将文本输出，并用管道“|”将其作为脚本 mapper.py 的标准输入，执行脚本 mapper.py，查看 Map 阶段的输出。

```
cat test.txt | ./mapper.py
deer        1
deer        1
river       1
car         1
river       1
deer        1
```

进一步加入管道“|”，并使用 sort 命令按字母排序，查看 Reduce 阶段的输入。

```
cat test.txt | ./mapper.py | sort
car         1
deer        1
deer        1
deer        1
river       1
river       1
```

进一步加入管道“|”，并执行脚本 reducer.py，查看 Reduce 阶段的输出。

```
cat test.txt | ./mapper.py | sort | ./reducer.py
car         1
deer        3
river       2
```

4. 数据准备

使用 hdfs dfs -put 命令从本地文件系统将文件夹 data/novels 上传到 HDFS。

```
hdfs dfs -put data/novels/
```

5. 提交到 Hadoop 集群执行

使用 hadoop jar 命令将 MapReduce 程序提交到 Hadoop 集群执行，其中：

- 第 1 个参数表示程序 jar 包，这里指定 Hadoop 流处理包；
- -files 参数表示程序执行需要的文件列表；
- -mapper 参数表示 map( )函数脚本；
- -reducer 参数表示 reduce( )函数脚本；
- -input 参数表示输入数据路径；
- -output 参数表示输出数据路径。

```
hadoop jar /usr/lib/hadoop-mapreduce/hadoop-streaming.jar -files mapper.py,reducer.py -mapper mapper.py -reducer reducer.py -input novels -output output
packageJobJar: [] [/usr/lib/hadoop-mapreduce/hadoop-streaming-3.0.0-cdh6.1.0.jar]/tmp/streamjob6402560227621776907.jar tmpDir=null
18/10/01 14:36:55 INFO client.RMProxy: Connecting to ResourceManager at /0.0.0.0:8032
18/10/01 14:36:56 INFO client.RMProxy: Connecting to ResourceManager at /0.0.0.0:8032
18/10/01 14:36:57 INFO mapred.FileInputFormat: Total input paths to process : 3
18/10/01 14:36:58 INFO mapreduce.JobSubmitter: number of splits:3
18/10/01 14:36:58 INFO mapreduce.JobSubmitter: Submitting tokens for job: job_1538355655973_0003
18/10/01 14:36:59 INFO impl.YarnClientImpl: Submitted application application_1538355655973_0003
18/10/01 14:36:59 INFO mapreduce.Job: The url to track the job: http://localhost:8088/proxy/application_1538355655973_0003/
18/10/01 14:36:59 INFO mapreduce.Job: Running job: job_1538355655973_ 0003
18/10/01 14:37:08 INFO mapreduce.Job: Job job_1538355655973_0003 running in uber mode : false
18/10/01 14:37:08 INFO mapreduce.Job: map 0% reduce 0%
18/10/01 14:37:17 INFO mapreduce.Job: map 33% reduce 0%
18/10/01 14:37:18 INFO mapreduce.Job: map 67% reduce 0%
18/10/01 14:37:23 INFO mapreduce.Job: map 100% reduce 0%
18/10/01 14:37:26 INFO mapreduce.Job: map 100% reduce 100%
18/10/01 14:37:27 INFO mapreduce.Job: Job job_1538355655973_0003 completed successfully
18/10/01 14:37:28 INFO mapreduce.Job: Counters: 49
    File System Counters
        FILE: Number of bytes read=3441952
        FILE: Number of bytes written=7472445
        FILE: Number of read operations=0
        FILE: Number of large read operations=0
        FILE: Number of write operations=0
        HDFS: Number of bytes read=2072252
        HDFS: Number of bytes written=158322
```

```
        HDFS: Number of read operations=12
        HDFS: Number of large read operations=0
        HDFS: Number of write operations=2
    Job Counters
        Launched map tasks=3
        Launched reduce tasks=1
        Data-local map tasks=3
        Total time spent by all maps in occupied slots (ms)=37292
        Total time spent by all reduces in occupied slots (ms)=13090
        Total time spent by all map tasks (ms)=18646
        Total time spent by all reduce tasks (ms)=6545
        Total vcore-milliseconds taken by all map tasks=18646
        Total vcore-milliseconds taken by all reduce tasks=6545
        Total megabyte-milliseconds taken by all map tasks=9546752
        Total megabyte-milliseconds taken by all reduce tasks=3351040
    Map-Reduce Framework
        Map input records=28437
        Map output records=371964
        Map output bytes=2698018
        Map output materialized bytes=3441964
        Input split bytes=338
        Combine input records=0
        Combine output records=0
        Reduce input groups=14951
        Reduce shuffle bytes=3441964
        Reduce input records=371964
        Reduce output records=14951
        Spilled Records=743928
        Shuffled Maps =3
        Failed Shuffles=0
        Merged Map outputs=3
        GC time elapsed (ms)=657
        CPU time spent (ms)=10480
        Physical memory (bytes) snapshot=984510464
        Virtual memory (bytes) snapshot=9444147200
        Total committed heap usage (bytes)=799539200
    Shuffle Errors
        BAD_ID=0
        CONNECTION=0
        IO_ERROR=0
        WRONG_LENGTH=0
        WRONG_MAP=0
        WRONG_REDUCE=0
    File Input Format Counters
        Bytes Read=2071914
    File Output Format Counters
        Bytes Written=158322
18/10/01 14:37:28 INFO streaming.StreamJob: Output directory: output
```

可以看出，输出信息包含了丰富的内容。

在作业开始执行前，输出的信息包括：

- mapred.FileInputFormat: Total input paths to process 表示输入文件数量为 3；

- mapreduce.JobSubmitter: number of splits 表示输入文件的划分数量为 3，这里即每个文件只有一个划分；
- mapreduce.JobSubmitter: Submitting tokens for job 表示作业的标识符为 job_1538355655973_0003，查看状态或杀死作业时使用；
- mapreduce.Job: The url to track the job 表示跟踪作业运行状态的 URL 地址，可以在浏览器中打开。

在作业执行过程中，会实时输出类似 map X% reduce X%的进度情况。

在作业执行完毕后，输出多个计数器，包括：

- File System Counters 表示文件系统计数器，其中又包括：
  - FILE: Number of bytes read/written 表示操作系统读取/写入的文件字节数；
  - HDFS: Number of bytes read/written 表示 HDFS 读取/写入文件字节数。
- Job Counters 表示作业计数器，其中又包括：
  - Launched map/reduce tasks 表示 Map/Reduce 任务数；
  - Total time spent by all map/reduce tasks (ms)表示所有 Map/Reduce 任务花费的毫秒数。
- Map-Reduce Framework 表示 MapReduce 框架计数器，其中又包括：
  - Map/Combine/Reduce input records 表示 Map/Combine/Reduce 任务输入记录数；
  - Map/Combine/Reduce output records 表示 Map/Combine/Reduce 任务输出记录数；
  - Map output bytes 表示 Map 任务输出字节数；
  - Map output materialized bytes 表示 Map 任务输出且在写入硬盘的字节数，Map 任务的输出作为中间结果会写入任务所在的操作系统硬盘；
  - Reduce input groups 表示 Reduce 任务的输入分组数；
  - Reduce shuffle bytes 表示 Shuffle 操作的字节数。

6. 查看结果

使用 hdfs dfs -ls 命令显示文件（夹）统计信息。

```
hdfs dfs -ls output
Found 2 items
-rw-r--r--    1 root hadoop           0 2018-10-01 14:37 output/_SUCCESS
-rw-r--r--    1 root hadoop      158322 2018-10-01 14:37 output/part-00000
```

可以看出，输出文件夹中包含 2 个文件，文件_SUCCESS 是一个空文件，大小为 0，仅表示作业执行成功，文件 part-00000 是输出数据文件，其中包含了词频统计结果。

使用 hdfs dfs -tail 命令显示文件末尾 1 KB 内容。

```
hdfs dfs -tail output/part-00000
wouldn    44
wound     10
wove      4
woven     1
wrap      3
[...]
```

4.5 MapReduce 程序的 Reducer 数量

前面已经提到，Reduce 任务的数量并不取决于输入数据大小，因为在 MapReduce 作业启动前，框架无法预知 Map 任务的输出数据大小。 Reduce 任务的数量通常情况下由用户或更高层的应用指定。

Hadoop 流处理命令中，指定选项 numReduceTasks 表示 Reducer 数量，这里设为 3。

```
hadoop jar /usr/lib/hadoop-mapreduce/hadoop-streaming.jar -files mapper.py,reducer.py -mapper mapper.py -reducer reducer.py -input novels -output output1 -numReduceTasks 3
[...]
18/10/01 14:45:15 INFO mapreduce.Job: Counters: 50
    File System Counters
        FILE: Number of bytes read=3441964
        FILE: Number of bytes written=7766801
        FILE: Number of read operations=0
        FILE: Number of large read operations=0
        FILE: Number of write operations=0
        HDFS: Number of bytes read=2072252
        HDFS: Number of bytes written=158322
        HDFS: Number of read operations=18
        HDFS: Number of large read operations=0
        HDFS: Number of write operations=6
    Job Counters
        Killed map tasks=1
        Launched map tasks=3
        Launched reduce tasks=3
        Data-local map tasks=3
        Total time spent by all maps in occupied slots (ms)=35878
        Total time spent by all reduces in occupied slots (ms)=36218
        Total time spent by all map tasks (ms)=17939
        Total time spent by all reduce tasks (ms)=18109
        Total vcore-milliseconds taken by all map tasks=17939
        Total vcore-milliseconds taken by all reduce tasks=18109
        Total megabyte-milliseconds taken by all map tasks=9184768
        Total megabyte-milliseconds taken by all reduce tasks= 9271808
    Map-Reduce Framework
        Map input records=28437
        Map output records=371964
        Map output bytes=2698018
        Map output materialized bytes=3442000
        Input split bytes=338
        Combine input records=0
        Combine output records=0
        Reduce input groups=14951
        Reduce shuffle bytes=3442000
        Reduce input records=371964
        Reduce output records=14951
        Spilled Records=743928
        Shuffled Maps =9
```

```
        Failed Shuffles=0
        Merged Map outputs=9
        GC time elapsed (ms)=731
        CPU time spent (ms)=15560
        Physical memory (bytes) snapshot=1376632832
        Virtual memory (bytes) snapshot=14196301824
        Total committed heap usage (bytes)=1156055040
    Shuffle Errors
        BAD_ID=0
        CONNECTION=0
        IO_ERROR=0
        WRONG_LENGTH=0
        WRONG_MAP=0
        WRONG_REDUCE=0
    File Input Format Counters
        Bytes Read=2071914
    File Output Format Counters
        Bytes Written=158322
18/10/01 14:45:15 INFO streaming.StreamJob: Output directory: output1
```

使用命令 hdfs dfs -ls 显示文件（夹）统计信息。

```
hdfs dfs -ls output1
Found 4 items
-rw-r--r--    1 root hadoop          0 2018-10-01 14:45 output1/_SUCCESS
-rw-r--r--    1 root hadoop      52341 2018-10-01 14:45 output1/part-00000
-rw-r--r--    1 root hadoop      53503 2018-10-01 14:45 output1/part-00001
-rw-r--r--    1 root hadoop      52478 2018-10-01 14:45 output1/part-00002
```

可以看出，除了文件_SUCCESS 以外，另有 3 个以 part-开头的输出数据文件，每个文件对应一个 Reducer 的输出。

4.6 MapReduce 程序的 Combiner

Combiner 又称迷你版 Reducer，用于在 Mapper 的输出通过网络传输到 Reduce 任务前，根据键合并 Mapper 的输出，以减少网络传输。Combiner 与 Mapper 一一对应，即如果配置启用 Combiner，则每个 Mapper 会对应一个 Combiner，并且该 Combiner 只会接受来自所对应的 Mapper 的输出。

考虑：在词频统计例子中，有 2 个 Mapper 任务，其中第一个 Mapper 的输入为 deer deer river，而另一个 Mapper 的输入为 car car。

当没有启用 Combiner 时，第一个 Mapper 输出到 Reducer 的记录为 3 条，即：

```
(deer, 1)
(deer, 1)
(river, 1)
```

另一个 Mapper 输出到 Reducer 的记录为 2 条，即：

```
(car, 1)
(car, 1)
```

当启用 Combiner 时，第一个 Mapper 输出经 Combiner 合并，输入到 Reducer 的记录为 2 条，即：

```
(deer, 2)
(river, 1)
```

另一个 Mapper 输出经 Combiner 合并，输入到 Reducer 的记录为 1 条，即：

```
(car, 2)
```

因此，通过启用 Combiner 节省了网络传输。

Hadoop 流处理命令中，指定选项 combiner 表示 Combiner 函数脚本。

```
hadoop jar /usr/lib/hadoop-mapreduce/hadoop-streaming.jar -files mapper.py,reducer.py -mapper mapper.py
-reducer reducer.py -input novels -output output2 -numReduceTasks 3 -combiner reducer.py
18/10/01 14:49:28 INFO mapreduce.Job: Counters: 50
    File System Counters
        FILE: Number of bytes read=307141
        FILE: Number of bytes written=1499279
        FILE: Number of read operations=0
        FILE: Number of large read operations=0
        FILE: Number of write operations=0
        HDFS: Number of bytes read=2072252
        HDFS: Number of bytes written=158322
        HDFS: Number of read operations=18
        HDFS: Number of large read operations=0
        HDFS: Number of write operations=6
    Job Counters
        Killed reduce tasks=1
        Launched map tasks=3
        Launched reduce tasks=3
        Data-local map tasks=3
        Total time spent by all maps in occupied slots (ms)=53650
        Total time spent by all reduces in occupied slots (ms)=44242
        Total time spent by all map tasks (ms)=26825
        Total time spent by all reduce tasks (ms)=22121
        Total vcore-milliseconds taken by all map tasks=26825
        Total vcore-milliseconds taken by all reduce tasks=22121
        Total megabyte-milliseconds taken by all map tasks=13734400
        Total megabyte-milliseconds taken by all reduce tasks= 11325952
    Map-Reduce Framework
        Map input records=28437
        Map output records=371964
        Map output bytes=2698018
        Map output materialized bytes=307177
        Input split bytes=338
        Combine input records=371964
        Combine output records=25093
        Reduce input groups=14951
        Reduce shuffle bytes=307177
        Reduce input records=25093
```

```
        Reduce output records=14951
        Spilled Records=50186
        Shuffled Maps =9
        Failed Shuffles=0
        Merged Map outputs=9
        GC time elapsed (ms)=2603
        CPU time spent (ms)=15970
        Physical memory (bytes) snapshot=1638469632
        Virtual memory (bytes) snapshot=14183243776
        Total committed heap usage (bytes)=1357381632
    Shuffle Errors
        BAD_ID=0
        CONNECTION=0
        IO_ERROR=0
        WRONG_LENGTH=0
        WRONG_MAP=0
        WRONG_REDUCE=0
    File Input Format Counters
        Bytes Read=2071914
    File Output Format Counters
        Bytes Written=158322
18/10/01 14:49:28 INFO streaming.StreamJob: Output directory: output2
```

与上一节没有 Reducer 对比，Shuffle 操作的字节数从 3 442 000 减少到 307 177，减少了一个数量级。

4.7 MapReduce 程序实现数据连接

数据连接是指将两个数据集按照某些条件（通常为某一列的值相等）合并成一个数据集。

本节的例子基于两个数据集。第一个数据集是 2014 年前 10 个月由纽约出发航班的准点情况数据，由美国运输局收集整理，包含 253 316 个样本，每个样本包含 17 个属性，如表 4-1 所示。

表 4-1　航班准点情况数据集属性定义

属　性	定　义	属　性	定　义
year	年份	tailnum	尾翼编号
month	月份	flight	航班编号
day	日期	origin	出发地
dep_time	出发时间	dest	目的地
dep_delay	出发延误分钟数	air_time	飞行分钟数
arr_time	到达时间	distance	距离
arr_delay	到达延误分钟数	hour	小时
carrier	航空公司代码	min	分钟

该数据集的路径为～/data/flights/flights14.csv。

```
head ~/data/flights/flights14.csv
year,month,day,dep_time,dep_delay,arr_time,arr_delay,cancelled,carrier,tailnum,flight,origin,dest,air_ti
me,distance,hour,min
2014,1,1,914,14,1238,13,0,AA,N338AA,1,JFK,LAX,359,2475,9,14
2014,1,1,1157,-3,1523,13,0,AA,N335AA,3,JFK,LAX,363,2475,11,57
2014,1,1,1902,2,2224,9,0,AA,N327AA,21,JFK,LAX,351,2475,19,2
2014,1,1,722,-8,1014,-26,0,AA,N3EHAA,29,LGA,PBI,157,1035,7,22
2014,1,1,1347,2,1706,1,0,AA,N319AA,117,JFK,LAX,350,2475,13,47
2014,1,1,1824,4,2145,0,0,AA,N3DEAA,119,EWR,LAX,339,2454,18,24
2014,1,1,2133,-2,37,-18,0,AA,N323AA,185,JFK,LAX,338,2475,21,33
2014,1,1,1542,-3,1906,-14,0,AA,N328AA,133,JFK,LAX,356,2475,15,42
2014,1,1,1509,-1,1828,-17,0,AA,N5FJAA,145,JFK,MIA,161,1089,15,9
```

第二个数据集是航空公司简写与描述的映射数据集，由美国运输局收集整理，包含 1 889 个样本，每个样本包含 2 个属性，如表 4-2 所示。

表 4-2　航空公司简写与描述的映射数据集属性定义

属　性	定　义	属　性	定　义
carrier	航空公司代码	description	描述

该数据集的路径为～/data/flights/carrier.csv。

```
head ~/data/flights/carrier.csv
carrier,description
"02Q","Titan Airways (2006 - )"
"04Q","Tradewind Aviation (2006 - )"
"06Q","Master Top Linhas Aereas Ltd. (2007 - )"
"07Q","Flair Airlines Ltd. (2007 - )"
"09Q","Swift Air, LLC d/b/a Eastern Air Lines d/b/a Eastern (2018 - )"
"0BQ","DCA (2007 - )"
"0CQ","ACM AIR CHARTER GmbH (2007 - )"
"0FQ","Maine Aviation Aircraft Charter, LLC (2017 - )"
"0HQ","Polar Airlines de Mexico d/b/a Nova Air (2007 - )"
```

可以看出，两个数据文件的第一行都是数据的变量名，且数据集的各变量都以逗号分隔。第二个数据集 carriers.csv 中的每个变量内容都使用了双引号"。这两个数据集的公共变量是航空公司代码（变量 carrier），即第一个数据集的第 9 列和第二个数据集的第 1 列。以下例子将以该变量相等作为连接条件。

1. map( )函数

创建 map( )函数的 Python 脚本 join_mapper.py。脚本读取标准输入，遍历每行文本。调用 strip( )函数去除字符串首尾不可见字符，并调用 split( )函数将每行文本按逗号做分隔。由于输入的数据文件可能来自包含 17 个变量的航班准点情况数据，也可能来自包含 2 个变量的航空公司简写与描述映射数据，因此需要进行区分处理。如果分隔后有 17 个元素，则表示该行数据来自航班准点情况数据，将第 9 个元素作为键，整行内容作为值，并添加一个来源标记 flight，用于在 reduce( )函数中区分数据来自哪个数据集。如果隔后没有 17 个元素，则表示该行数据来自航空公司简写与描述映射数据，将第 1 个元素作为键并调用 replace( )函数去除双引号，键后面的内容作为值，

并添加一个来源标记 carrier。最终以制表符做分隔，调用 print( )函数将数据连接键、来源标记和值写入标准输出。

```
#!/usr/bin/env python
import sys, re
for line in sys.stdin:
    line = line.strip( )
    fields = line.split(",")
    if len(fields) == 17:
        key = fields[8]
        source = "flight"
        value = line
    else:
        key = fields[0].replace("\"", "")
        source = "carrier"
        value = line.split(",", 1)[1]
    if (key != "year") & (key != "carrier"):
        print( "%s\t%s\t%s" % (key, source, value) )
```

2．reduce( )函数

创建 reduce( )函数的 Python 脚本 join_reducer.py。脚本首先定义变量 last_key 存储上一行的键。读取标准输入，遍历每行文本。调用 strip( )函数去除字符串首尾不可见字符，再调用函数 split( )将每一行文本按制表符分隔成数据连接键、来源标记和值。由于 Reduce 任务的输入都已经做了排序，且对于同样的数据连接键，来源标记 carrier 排在 flight 前面（按字母顺序），如果当前行的键与上一行的键不相同，则将航空公司描述（变量 carrier_description）设为空字符串。如果该行的来源标记表示来自航空公司简写与描述映射数据，则将值赋给变量 carrier_description，否则表示来自航班准点情况数据，则调用函数 print( )将值和变量 carrier_description 写入标准输出。

```
#!/usr/bin/env python
import sys
last_key = None
for input_line in sys.stdin:
    input_line = input_line.strip( )
    this_key, source, value = input_line.split("\t", 2)
    if this_key != last_key:
        carrier_description = ""
        last_key = this_key
    if source == "carrier":
        carrier_description = value
    else:
        print( "%s,%s" % (value, carrier_description) )
```

3．本地测试

创建一个测试文本文件 flights14_sample.csv，为航班准点情况数据的一小部分。

```
2014,1,1,1754,24,2041,16,0,AA,N623AA,2448,JFK,EGE,266,1746,17,54
2014,1,1,647,-3,942,-13,0,AA,N5CEAA,2493,JFK,MCO,147,944,6,47
2014,1,1,1829,4,2140,-12,0,AS,N423AS,5,EWR,SEA,347,2402,18,29
2014,1,1,709,4,1030,15,0,AS,N435AS,15,EWR,SEA,348,2402,7,9
```

```
2014,1,1,846,51,1100,60,0,B6,N318JB,1273,JFK,CHS,109,636,8,46
2014,1,1,1353,41,1657,51,0,B6,N318JB,553,JFK,PBI,165,1028,13,53
2014,1,1,2127,72,2215,49,0,B6,N318JB,418,JFK,BOS,32,187,21,27
```

创建另一个测试文本文件 carrier_sample.csv，为航空公司简写与描述映射数据的一小部分。

```
"AA","American Airlines Inc. (1960 - )"
"B6","JetBlue Airways (2000 - )"
"CA","Air China (1981 - )"
```

在提交到 Hadoop 执行前，先在本地进行测试。

使用 chmod 命令开放脚本的执行权限。

```
chmod 755 join_mapper.py
chmod 755 join_reducer.py
```

使用 cat 命令将文件 flights14_sample.csv 输出，并用管道“|”将其作为脚本 join_mapper.py 的标准输入，执行脚本 join_mapper.py，查看 Map 阶段的输出。

```
cat flights14_sample.csv | ./join_mapper.py
AA    flight2014,1,1,1754,24,2041,16,0,AA,N623AA,2448,JFK,EGE, 266, 1746,17,54
AA    flight2014,1,1,647,-3,942,-13,0,AA,N5CEAA,2493,JFK,MCO, 147, 944,6,47
AS    flight2014,1,1,1829,4,2140,-12,0,AS,N423AS,5,EWR,SEA, 347, 2402,18,29
AS    flight2014,1,1,709,4,1030,15,0,AS,N435AS,15,EWR,SEA, 348, 2402,7,9
B6    flight2014,1,1,846,51,1100,60,0,B6,N318JB,1273,JFK,CHS, 109,636,8,46
B6    flight2014,1,1,1353,41,1657,51,0,B6,N318JB,553,JFK,PBI,165, 1028,13,53
B6    flight2014,1,1,2127,72,2215,49,0,B6,N318JB,418,JFK,BOS,32, 187,21,27
```

对文件 carrier_sample.csv 进行同样操作。

```
cat carrier_sample.csv | ./join_mapper.py
AA    carrier "American Airlines Inc. (1960 - )"
B6    carrier "JetBlue Airways (2000 - )"
CA    carrier "Air China (1981 - )"
```

进一步加入管道“|”，并使用 sort 命令按字母排序，查看 Reduce 阶段的输入。

```
cat *.csv | ./join_mapper.py | sort
AA    carrier "American Airlines Inc. (1960 - )"
AA    flight    2014,1,1,1754,24,2041,16,0,AA,N623AA,2448,JFK,EGE,266,1746,17,54
AA    flight    2014,1,1,647,-3,942,-13,0,AA,N5CEAA,2493,JFK,MCO, 147,944,6,47
AS    flight    2014,1,1,1829,4,2140,-12,0,AS,N423AS,5,EWR,SEA,347, 2402,18,29
AS    flight    2014,1,1,709,4,1030,15,0,AS,N435AS,15,EWR,SEA,348, 2402,7,9
B6    carrier "JetBlue Airways (2000 - )"
B6    flight    2014,1,1,1353,41,1657,51,0,B6,N318JB,553,JFK,PBI,165, 1028,13,53
B6    flight    2014,1,1,2127,72,2215,49,0,B6,N318JB,418,JFK,BOS,32, 187,21,27
B6    flight    2014,1,1,846,51,1100,60,0,B6,N318JB,1273,JFK,CHS, 109,636,8,46
CA    carrier "Air China (1981 - )"
```

进一步加入管道“|”，并执行脚本 join_reducer.py，查看 Reduce 阶段的输出。

```
cat *.csv | ./join_mapper.py | sort | ./join_reducer.py
2014,1,1,1754,24,2041,16,0,AA,N623AA,2448,JFK,EGE,266,1746,17,54,"American Airlines Inc.(1960 - )"
```

```
2014,1,1,647,-3,942,-13,0,AA,N5CEAA,2493,JFK,MCO,147,944,6,47,"American Airlines Inc.(1960 - )"
2014,1,1,1829,4,2140,-12,0,AS,N423AS,5,EWR,SEA,347,2402,18,29,
2014,1,1,709,4,1030,15,0,AS,N435AS,15,EWR,SEA,348,2402,7,9,
2014,1,1,1353,41,1657,51,0,B6,N318JB,553,JFK,PBI,165,1028,13,53, "JetBlue Airways (2000 - )"
2014,1,1,2127,72,2215,49,0,B6,N318JB,418,JFK,BOS,32,187,21,27,"JetBlue Airways (2000 - )"
2014,1,1,846,51,1100,60,0,B6,N318JB,1273,JFK,CHS,109,636,8,46,"JetBlue Airways (2000 - )"
```

4. 数据准备

使用 hdfs dfs -put 命令从本地文件系统将文件夹 data/flights 上传到 HDFS。

```
hdfs dfs -put data/flights/
```

5. 提交到 Hadoop 集群执行

使用 hadoop jar 命令将 MapReduce 程序提交到 Hadoop 集群执行。

```
hadoop jar/usr/lib/hadoop-mapreduce/hadoop-streaming.jar -files join_mapper.py,join_reducer.py -mapper
join_mapper.py -reducer join_reducer.py -input flights -output output3
packageJobJar: [] [/usr/lib/hadoop-mapreduce/hadoop-streaming-3.0.0-cdh6.1.0.jar] /tmp/streamjob4918
10231080136661.jar tmpDir=null
18/10/08 10:47:14 INFO client.RMProxy: Connecting to ResourceManager at /0.0.0.0:8032
18/10/08 10:47:14 INFO client.RMProxy: Connecting to ResourceManager at /0.0.0.0:8032
18/10/08 10:47:16 INFO mapred.FileInputFormat: Total input paths to process : 2
18/10/08 10:47:16 INFO mapreduce.JobSubmitter: number of splits:3
18/10/08 10:47:16 INFO mapreduce.JobSubmitter: Submitting tokens for job: job_1538708448332_0001
18/10/08 10:47:17 INFO impl.YarnClientImpl: Submitted application application_1538708448332_0001
18/10/08 10:47:17 INFO mapreduce.Job: The url to track the job: http://localhost:8088/proxy/application_
1538708448332_0001/
18/10/08 10:47:17 INFO mapreduce.Job: Running job: job_1538708448332_0001
18/10/08 10:47:25 INFO mapreduce.Job: Job job_1538708448332_0001 running in uber mode : false
18/10/08 10:47:25 INFO mapreduce.Job: map 0% reduce 0%
18/10/08 10:47:36 INFO mapreduce.Job: map 67% reduce 0%
18/10/08 10:47:47 INFO mapreduce.Job: map 100% reduce 0%
18/10/08 10:47:50 INFO mapreduce.Job: map 100% reduce 100%
18/10/08 10:47:51 INFO mapreduce.Job: Job job_1538708448332_0001 completed successfully
18/10/08 10:47:51 INFO mapreduce.Job: Counters: 49
    File System Counters
        FILE: Number of bytes read=19211149
        FILE: Number of bytes written=39011063
        FILE: Number of read operations=0
        FILE: Number of large read operations=0
        FILE: Number of write operations=0
        HDFS: Number of bytes read=16172938
        HDFS: Number of bytes written=24015343
        HDFS: Number of read operations=12
        HDFS: Number of large read operations=0
        HDFS: Number of write operations=2
    Job Counters
        Launched map tasks=3
        Launched reduce tasks=1
        Data-local map tasks=3
```

```
		Total time spent by all maps in occupied slots (ms)=47526
		Total time spent by all reduces in occupied slots (ms)=23576
		Total time spent by all map tasks (ms)=23763
		Total time spent by all reduce tasks (ms)=11788
		Total vcore-milliseconds taken by all map tasks=23763
		Total vcore-milliseconds taken by all reduce tasks=11788
		Total megabyte-milliseconds taken by all map tasks=12166656
		Total megabyte-milliseconds taken by all reduce tasks= 6035456
	Map-Reduce Framework
		Map input records=253742
		Map output records=253740
		Map output bytes=18703663
		Map output materialized bytes=19211161
		Input split bytes=313
		Combine input records=0
		Combine output records=0
		Reduce input groups=426
		Reduce shuffle bytes=19211161
		Reduce input records=253740
		Reduce output records=253316
		Spilled Records=507480
		Shuffled Maps =3
		Failed Shuffles=0
		Merged Map outputs=3
		GC time elapsed (ms)=384
		CPU time spent (ms)=7260
		Physical memory (bytes) snapshot=1159254016
		Virtual memory (bytes) snapshot=9453387776
		Total committed heap usage (bytes)=1009778688
	Shuffle Errors
		BAD_ID=0
		CONNECTION=0
		IO_ERROR=0
		WRONG_LENGTH=0
		WRONG_MAP=0
		WRONG_REDUCE=0
	File Input Format Counters
		Bytes Read=16172625
	File Output Format Counters
		Bytes Written=24015343
18/10/08 10:47:51 INFO streaming.StreamJob: Output directory: output
```

6．查看结果

使用 hdfs dfs –ls 命令显示文件（夹）统计信息。

```
hdfs dfs -ls output
Found 2 items
-rw-r--r--    1 root hadoop           0 2018-10-08 10:47 output/_SUCCESS
-rw-r--r--    1 root hadoop    24015343 2018-10-08 10:47 output/part-00000
```

使用 hdfs dfs –tail 命令显示文件末尾 1 KB 内容。

```
hdfs dfs -tail output/part-00000
2014,9,11,850,-5,1012,-13,0,WN,N710SW,3415,LGA,MDW,118,725,8,50, "Southwest Airlines Co. (1979 - )"
2014,9,11,555,-5,707,-3,0,WN,N8327A,4963,LGA,MDW,109,725,5,55,"Southwest Airlines Co. (1979 - )"
2014,9,11,925,-5,1046,-24,0,WN,N555LV,786,LGA,MKE,121,738,9,25,"Southwest Airlines Co. (1979 - )"
2014,9,11,558,-2,718,-2,0,WN,N957WN,1027,LGA,MKE,120,738,5,58,"Southwest Airlines Co. (1979 - )"
2014,9,11,1404,-6,1515,-20,0,WN,N298WN,1345,LGA,MKE,118,738,14,4,"Southwest Airlines Co. (1979 - )"
2014,9,11,1931,1,2049,-11,0,WN,N7715E,2164,LGA,MKE,123,738,19,31,"Southwest Airlines Co. (1979 - )"
2014,9,11,1824,-1,1955,-10,0,WN,N236WN,791,LGA,STL,132,888,18,24,"Southwest Airlines Co. (1979 - )"
2014,9,11,1154,-6,1333,-22,0,WN,N467WN,2414,LGA,STL,134,888,11,54,"Southwest Airlines Co. (1979 - )"
2014,9,11,710,-5,846,-9,0,WN,N772SW,3755,LGA,STL,136,888,7,10,"Southwest Airlines Co. (1979 - )"
2014,10,31,1310,0,1600,-30,0,WN,N231WN,244,EWR,AUS,217,1504,13,10,"Southwest Airlines Co. (1979 - )"
```

小　　结

本章主要介绍了 MapReduce 的工作原理，包括 Map 阶段和 Reduce 阶段，以及中间的 Shuffle 阶段。MapReduce 程序还可以进一步配置 Reducer 数量和 Combiner 等，都会对数据处理过程产生影响。

本章还举了两个 MapReduce 程序的例子，第一个是实现词频统计，第二个是实现数据连接。其基本流程都包括定义 map( )函数、定义 reduce( )函数、本地测试、数据准备、提交到 Hadoop 集群执行和查看结果。

习　　题

1. MapReduce 程序的工作原理和处理流程是什么？
2. MapReduce 程序的 Reducer 数量对输出结果有什么影响？
3. MapReduce 程序的 Combiner 对中间结果有什么影响？
4. Hadoop 流处理的原理是什么？
5. 编写 MapReduce 程序统计各首字母的词频。
6. 编写 MapReduce 程序统计航班数据中每个月份记录数。

Hive 数据仓库

第 5 章 Hive 简介

5.1 Hive 概述

Hive 是一个基于 Hadoop 的数据仓库软件，可以使用类似 SQL 的语法读写、查询和管理大数据集。Hive 最初只是 Facebook 内部的一个项目，使得拥有 SQL 技能但缺乏 Java 编程技能的分析师能够查询存储在 HDF 上的海量数据。如今，Hive 已经成为一个成功的 Apache 项目，许多公司将它用于通用的、可扩展的数据处理平台上。

Hive 主要提供以下特性：

- 通过 SQL 访问数据以及常用的数据仓库操作如 ETL（Extract、Transform、Load）、报表和数据分析；
- 将数据结构施加于多种数据格式（如 CSV/TSV 文本文件、Parquet 和 ORC）的机制，数据可以存储在 HDFS 或 HBase 中；
- 利用 MapReduce、Tez 或 Spark 作为查询的执行引擎，本书主要介绍最常用的基于 MapReduce 执行引擎。

5.1.1 Hive 与传统数据库

Hive 在很大程度上与传统关系型数据库非常类似，然而由于 Hive 是基于 HDFS 和 MapReduce 进行数据存储和处理，这深刻影响着 Hive 所支持的特性。随着 Hive 的发展，这些架构上的限制已经逐渐解除，使得 Hive 与传统数据库更趋类似。

1．读时模式和写时模式

模式（schema）指的是数据库的表结构，即数据表中包含的列以及相应的数据类型。在传统数据库中，数据在导入表的时候就会验证模式，只有在模式符合时才能成功导入。这种在数据写入时验证模式的方式称为写时模式（schema on write）。Hive 采取的是另一种方式，即在数据导入时并不验证模式，而只在查询时才验证，称为读时模式（schema on read）。

这两种方式各有利弊。读时模式在数据导入时，不需要将输入数据进行读取、解析以及序列化成数据库的内部存储格式。整个数据导入过程只是将数据文件复制或移动到指定位置。同时，读时模式也更灵活，对于同一个数据文件，可以施加多种不同的模式，用于不同的分析任务。Hive 中的外部表即支持这一特性，后续章节会具体介绍。而写时模式的数据查询效率较高，因为数据库可以在一些列上做索引，并在数据存储格式上做优化。然而，写时模式的缺点是数据导入的时间较长。

2. 修改、事务和索引

修改、事务和索引是传统数据库的标配。直到近几年，Hive 才开始支持这些特性，早先的 Hive 版本都不支持这些特性。这都是由于 Hive 基于 HDFS 和 MapReduce 进行数据存储和处理，HDFS 文件并不支持在文件任意位置做修改，修改数据表中的一条记录就必须将整个数据表变换为一张新的数据表。多数数据仓库应用在数据导入数据表后，并不需要修改数据，而大多为扫描数据表的全部或一部分。

Hive 很早之前就支持使用 INSERT INTO 语句将新的数据文件插入到数据表中。然而直到 0.14 版本才开始支持使用 INSERT INTO TABLE ... VALIES 语句插入小批量数据，同时也开始支持 UPDATE 和 DELETE 语句。HDFS 不支持在文件任意位置做修改，因此插入、修改和删除等操作的结果会保存在小的增量文件（delta file）中。后台的 MapReduce 作业会定期启动，将增量文件与基本表数据文件合并。这些插入、修改和删除操作都在事务（transaction）中完成。因此，对同一张数据表的不同查询能够看到一致的数据表。Hive 同时也支持事务锁。

索引可以加速数据查询。例如，查询 SELECT * FROM t WHERE x = a 可以利用列 x 上的索引，直接定位到相关的表数据文件位置，而不需要进行全表扫描。Hive 中目前支持两种索引类型：紧凑型（compact）和位图型（bitmap）。紧凑型索引存储每个值的 HDFS 文件块编号。位图型索引用位集合（bitset）存储数值对应的行号。

5.1.2 Hive 元数据存储

Hive 元数据存储（metastore）用于在关系型数据库（如 MySQL 或 Oracle）中存储 Hive 表和分区的元数据，并通过服务 API 将这些信息提供给客户端。

默认情况下，Hive 本身包含了一个内置的 Derby 数据库作为元数据存储。然而，该内置 Derby 数据库仅支持一个连接，意味着每次只有一个 Hive 会话可以访问元数据存储。这显然是生产环境中不能接受的。因此，多数情况下，会安装独立的关系型数据库（如 MySQL 或 Oracle）。

5.2 Hive 的安装

Hive 安装前需确保 Hadoop 已安装完成，具体流程参考之前章节。

1. 安装 Hive 相关组件（主节点）

使用 yum install 命令在线安装 Hive 核心组件，包括：

- HiveServer2 服务：hive-server2.noarch；
- 元数据存储服务：hive-metastore.noarch；
- Hive 客户端：hive。

```
sudo yum install hive-server2.noarch
sudo yum install hive-metastore.noarch
sudo yum install hive
```

2．配置基于 MySQL 的 Hive 元数据存储（主节点）

以 root 用户登录 MySQL。

```
mysql -u root -p
Enter password: 123456
```

使用 CREATE DATABASES 语句创建 Hive 元数据存储的数据库 metastore，并使用 USE 语句选择该数据库。

```
CREATE DATABASE metastore;
USE metastore;
```

使用 SOURCE 语句运行创建 Hive 元数据存储的相关数据表的 SQL 脚本/usr/lib/hive/scripts/metastore/upgrade/mysql/hive-schema-2.1.1.mysql.sql。

```
SOURCE /usr/lib/hive/scripts/metastore/upgrade/mysql/hive-schema-2.1.1.mysql.sql;
```

使用 GRANT 语句赋予 Hive 用户操作数据库 metastore 的所有权限，这里将 Hive 用户的密码设置为 123456。

```
GRANT ALL ON metastore.* TO'hive'@'%'IDENTIFIED BY'123456';
```

使用 SHOW DATABASES 和 SHOW GRANTS 验证数据库的创建和权限的赋予。

```
SHOW DATABASES;
+------------------------+
| Database               |
+------------------------+
| information_schema     |
| metastore              |
| mysql                  |
| performance_schema     |
+------------------------+
SHOW GRANTS FOR 'hive'@'%';
+------------------------------------------------------------------------------------------------------------------+
| Grants for hive@%                                                                                                |
+------------------------------------------------------------------------------------------------------------------+
| GRANT USAGE ON *.* TO 'hive'@'%' IDENTIFIED BY PASSWORD '*6BB4837EB74329105
EE4568DDA7DC67ED2CA2AD9'                                                                                         |
| GRANT ALL PRIVILEGES ON 'metastore'.* TO 'hive'@'%'                                                              |
+------------------------------------------------------------------------------------------------------------------+
```

使用 quit 命令退出 MySQL。

```
quit;
```

3．配置 MySQL JDBC 驱动器（主节点）

使用 ln 命令将 MySQL JDBC 驱动器文件 mysql-connector-java.jar 链接到文件夹/usr/lib/hive/lib/。

```
ln -s /usr/share/java/mysql-connector-java.jar /usr/lib/hive/lib/mysql-connector-java.jar
```

4．配置 Hive（主节点）

打开 Hive 配置文件/etc/hive/conf/hive-site.xml，加入的配置项包括：

- javax.jdo.option.ConnectionURL：Hive 元数据存储的链接地址；
- javax.jdo.option.ConnectionDriverName：Hive 元数据存储的驱动器名称，这里设置为 MySQL JDBC 驱动器；
- javax.jdo.option.ConnectionUserName：Hive 元数据存储的用户名；
- javax.jdo.option.ConnectionPassword：Hive 元数据存储的密码。

该文件配置部分（即<configuration>标签内）看起来如下所示。

```
vim /etc/hive/conf/hive-site.xml
<configuration>
    <property>
        <name>javax.jdo.option.ConnectionURL</name>
        <value>jdbc:mysql://master/metastore</value>
        <description>the URL of the MySQL database</description>
    </property>
    <property>
        <name>javax.jdo.option.ConnectionDriverName</name>
        <value>com.mysql.jdbc.Driver</value>
    </property>
    <property>
        <name>javax.jdo.option.ConnectionUserName</name>
        <value>hive</value>
    </property>
    <property>
        <name>javax.jdo.option.ConnectionPassword</name>
        <value>123456</value>
    </property>
</configuration>
```

5．创建 Hive 数据文件夹（主节点）

Hive 表中的数据通常存储在 HDFS 中的/user/hive/warehouse 文件夹中。

以 hdfs 用户使用 hdfs dfs -mkdir 命令创建文件夹/user/hive/warehouse 以及 hdfs dfs -chown 命令将文件夹所有权移交给 Hive 用户。

以 Hive 用户使用 hdfs dfs -chmod 命令将文件夹权限改为任何用户都可以读/写。

```
sudo -u hdfs hdfs dfs -mkdir -p /user/hive/warehouse
sudo -u hdfs hdfs dfs -chown -R hive /user/hive
sudo -u hive hdfs dfs -chmod 1777 /user/hive/warehouse
```

6．启动 Hive 相关组件（主节点）

使用 servimce start 命令启动 Hive 相关组件，包括：

- HiveServer2 服务：hive-server2；
- 元数据存储服务：hive-metastore。

```
sudo service hive-metastore start
sudo service hive-server2 start
Starting Hive Metastore (hive-metastore):                    [  OK  ]

Started Hive Server2 (hive-server2):                         [  OK  ]
```

7．验证 Hive 正常运行（主节点)

使用 beeline 命令执行显示所有数据表的命令。

```
beeline -u 'jdbc:hive2://master:10000/' -e 'select 1;'
Connecting to jdbc:hive2://master:10000/
Connected to: Apache Hive (version 2.1.1-cdh6.1.0)
Driver: Hive JDBC (version 2.1.1-cdh6.1.0)
Transaction isolation: TRANSACTION_REPEATABLE_READ
INFO  : Compiling command(queryId=hive_20190122120255_44a12068- 2a35-4b4b-b019-b233d0076773): select 1
INFO  : Semantic Analysis Completed
INFO  : Returning Hive schema: Schema(fieldSchemas:[FieldSchema (name:_c0, type:int, comment:null)], properties:null)
INFO  : Completed compiling command(queryId=hive_20190122120255_ 44a12068-2a35-4b4b-b019-b233d0076773); Time taken: 4.416 seconds
INFO  : Concurrency mode is disabled, not creating a lock manager
INFO  : Executing command(queryId=hive_20190122120255_44a12068- 2a35-4b4b-b019-b233d0076773): select 1
INFO  : Completed executing command(queryId=hive_20190122120255_ 44a12068-2a35-4b4b-b019-b233d0076773); Time taken: 0.003 seconds
INFO  : OK
+-------+
| _c0   |
+-------+
| 1     |
+-------+
1 row selected (5.329 seconds)
Beeline version 2.1.1-cdh6.1.0 by Apache Hive
```

5.3 Hive 的运行

Hive 中所有的查询都可以通过 Beeline 命令行客户端提交到 HiveServer2 服务，HiveServer2 服务会与元数据存储服务交互，检查权限并将查询映射到具体的 HDFS 文件，最终在 Hive 数据库中执行查询。整个过程如图 5-1 所示。

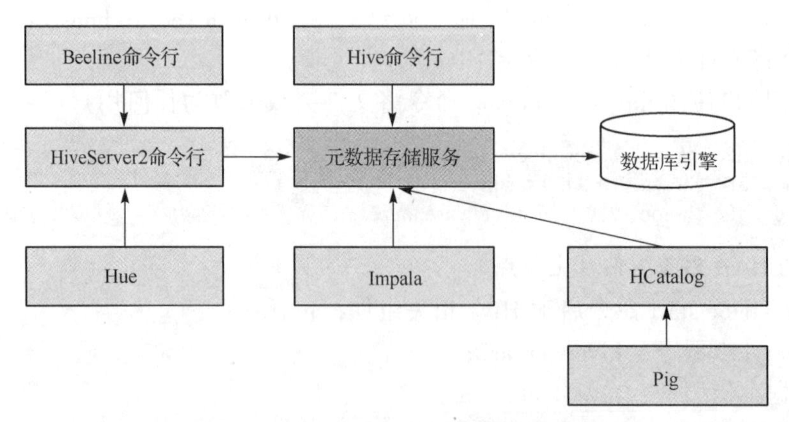

图 5-1 Hive 流程图

Hive 命令行从 Hive 1.0 后就已经被淘汰，转而使用 Beeline 作为命令行工具。Beeline 命令行和 Hue 与 HiveServer2 服务连接，将命令传递给元数据存储服务，并在 Hive 数据库中执行。而 Imapala 和 Pig 则直接连接元数据存储服务，而不通过 HiveServer2 服务。

1. Beeline 基本命令

使用 beeline 命令进入 Beeline 命令行。

```
beeline
Beeline version 2.1.1-cdh6.1.0 by Apache Hive
```

使用!connect jdbc:hive2://<HiveServer2 主机名>:<端口号>/<数据库名>命令连接到 Hive，并输入用户名（如 root）和密码（如 123456）。

```
!connect jdbc:hive2://master:10000/
Connecting to jdbc:hive2://master:10000/
Enter username for jdbc:hive2://master:10000/: root
Enter password for jdbc:hive2://master:10000/: 123456
Connected to: Apache Hive (version 2.1.1-cdh6.1.0)
Driver: Hive JDBC (version 2.1.1-cdh6.1.0)
Transaction isolation: TRANSACTION_REPEATABLE_READ
```

使用 set hivevar:<变量名>命令定义或显示变量值。

```
set hivevar:foo = bar;
set hivevar:foo;
+--------------------+--+
|       set          |
+--------------------+--+
| hivevar:foo=bar  |
+--------------------+--+
```

使用${hivevar:<变量名>}引用变量，Hive 会使用变量值替换掉变量引用。

```
select '${hivevar:foo}';
+------+--+
| _c0  |
+------+--+
| bar  |
+------+--+
```

使用 dfs <dfs 子命令>命令执行 HDFS 中的 dfs 命令，以分号结尾。

```
dfs -ls;
+---------------+--+
| DFS Output  |
+---------------+--+
+---------------+--+
```

使用!quit 命令退出 Beeline。

```
!quit
Closing: 0: jdbc:hive2://master:10000/
```

2. Beeline 运行选项

启动 Beeline 时指定一系列运行选项，可通过以下形式指定运行选项。

```
beeline -<选项 1> <值 1> -<选项 2> <值 2> …
```

常用的 Beeline 运行选项包括：

- -u：表示 HiveServer2 的 JDBC 连接 URL，形如 jdbc:hive2://<主机名>:<端口号>/<数据库名>;
- -n：表示登录用户名；
- -p：表示登录密码；
- -e：表示执行的查询语句，可以指定多次；
- -f：表示执行的脚本文件。

使用 beeline -u <JDBC 连接 URL> -n <登录用户名> -p <登录密码>命令启动 Beeline，避免每次启动都需要手动输入相关信息。

```
beeline -u 'jdbc:hive2://master:10000' -n root -p 123456
Beeline version 2.1.1-cdh6.1.0 by Apache Hive
!quit
```

使用 beeline -e <查询语句>命令以非交互的方式直接执行查询语句。

```
beeline -u 'jdbc:hive2://master:10000' -n root -p 123456 -e 'select 1;' -e 'select 2;'
Connecting to jdbc:hive2://master:10000
Connected to: Apache Hive (version 2.1.1-cdh6.1.0)
Driver: Hive JDBC (version 2.1.1-cdh6.1.0)
Transaction isolation: TRANSACTION_REPEATABLE_READ
[...]
INFO  : OK
+-------+--+
| _c0  |
+-------+--+
| 1     |
+-------+--+
1 row selected (0.347 seconds)
[...]
+------+--+
| _c0 |
+------+--+
| 2    |
+------+--+
1 row selected (0.087 seconds)
Beeline version 1.1.0-cdh5.15.1 by Apache Hive
Closing: 0: jdbc:hive2://master:10000
```

使用 beeline -f <脚本文件>命令以非交互的方式直接执行脚本文件。

```
echo 'select 1; select 2;' > script.sql
beeline -u 'jdbc:hive2://master:10000' -n root -p 123456 -f script.sql
Connecting to jdbc:hive2://master:10000
Connected to: Apache Hive (version 2.1.1-cdh6.1.0)
Driver: Hive JDBC (version 2.1.1-cdh6.1.0)
Transaction isolation: TRANSACTION_REPEATABLE_READ
```

```
0: jdbc:hive2://master:10000> select 1; select 2;
[...]
+------+--+
| _c0 |
+------+--+
| 1    |
+------+--+
1 row selected (0.334 seconds)
[...]
+------+--+
| _c0 |
+------+--+
| 2    |
+------+--+
1 row selected (0.082 seconds)
0: jdbc:hive2://master:10000>
0: jdbc:hive2://master:10000>
Closing: 0: jdbc:hive2://master:10000
```

小　结

本章首先介绍了 Hive 的产生背景和与传统关系型数据库的优势和劣势。Hive 是一个基于 Hadoop 的数据仓库软件，可以使用类似 SQL 的语法读/写、查询和管理大数据集。Hive 元数据存储用于在关系型数据库中存储 Hive 表和分区的元数据，并通过 API 服务将这些信息提供给客户端。

本章还着重介绍了 Hive 的安装流程，包括相关组件的安装（HiveServer2 服务、元数据存储服务和 Hive 客户端）、Hive 元数据存储的配置和验证安装结果。Hive 的客户端 Beeline 在运行时还可以指定多种选项控制运行行为。

习　题

1. Hive 与传统关系型数据库有什么区别？
2. Hive 元数据存储的作用是什么？
3. 熟悉 Hadoop 的安装流程，搭建 Hive 环境。
4. 验证 Hive 安装是否成功，并运行测试程序。
5. 熟悉 Hive 的常用运行选项，并逐一尝试。

第 6 章

Hive 数据定义 «

Hive 中的数据按照如下粒度从粗到细、从顶层到底层的方式组织。

① 数据库（database）：命名空间，为避免表、视图、分区和列等存在命名冲突，同时可以用作权限控制的单位。

② 数据表（table）：结构相同数据的存储单元。例如，对于用户网页浏览数据表，每一行数据可以包含以下列：

- 网页浏览的 UNIX 时间戳，为整数型；
- 用户标识符，为整数型；
- 网页 URL 地址，为字符串；
- 来源，即浏览的前一个网页 URL 地址，为字符串；
- 用户 IP 地址，为字符串。

③ 分区（partition）：每张表可以有一个或多个分区键，用于决定数据的存储方式。分区还能更高效地根据条件做行的筛选。例如，用户网页浏览数据表以国家和日期作为分区键，各分区键不同值的组合唯一地确定一个分区，所有用户国家为“中国”、日期为“2018-12-23”的数据组成了该表的一个分区。因此，如果仅需要在 2018 年 12 月 23 日的中国用户数据上做查询，则可以直接定位到该分区，大大加快了查询效率。

④ 桶（bucket）或簇（cluster）：分区中的数据可以根据某些列的哈希值进一步分成桶，可以用于高效地做数据抽样。例如，用户网页浏览数据表可以按用户标识符分桶。

需要注意的是，数据表并不一定要做分区或分桶，但是分区和分桶可以在数据查询时快速定位到感兴趣的数据，提高执行效率。

本章所有例子同样基于航班的准点情况数据，之前章节已有介绍。该数据集的路径为～/data/flights/flights14.csv，其中第一行为列名。同时也已经存储在 Hive 的 flights.flights14 表中。

6.1 数据库操作

1. 创建数据库

使用 CREATE DATABASE [IF NOT EXISTS] <数据库名>语句创建数据库 test1，其中：IF NOT EXISTS 关键词表示仅在该数据库不存在时才创建。

```
CREATE DATABASE IF NOT EXISTS test1;
```

使用 LOCATION 语句指定数据库在 HDFS 上的对应路径，默认为/user/hive/warehouse/<数据库名>.db，该默认路径可以在 hive-site.xml 配置文件中修改。这里创建数据库 test2，指定对应 HDFS 上的/user/hive/test2 文件夹。

```
CREATE DATABASE IF NOT EXISTS test2
LOCATION '/user/root/test2';
```

2．描述数据库

使用 DESCRIBE DATABASE <数据库名>语句描述数据库 test1，显示的信息包括数据库名称（db_name）、注释（comment）、存储位置（location）、所有者（owner_name）等。

```
DESCRIBE DATABASE test1;
+-----------+-----------+-----------------------------------+---------------+--------------+-------------+--+
| db_name  | comment |                location          | owner_name | owner_type | parameters |
+-----------+-----------+-----------------------------------+---------------+--------------+-------------+--+
| test1      |             | hdfs://master:8020/user/hive/warehouse/test1.db | root | USER     |              |
+-----------+-----------+-----------------------------------+---------------+--------------+-------------+--+
```

3．列出数据库

使用 SHOW DATABASES [LIKE '<正则表达式>']语句列出所有（满足正则表达式的）数据库。正则表达式中，星号*指代一个或多个字母，竖线|表示或。其中，default 数据库是 Hive 默认的数据库。

```
SHOW DATABASES;
+--------------------+--+
| database_name   |
+--------------------+--+
| default          |
| flights          |
| test1            |
| test2            |
+--------------------+--+
SHOW DATABASES LIKE 'test*';
+--------------------+--+
| database_name   |
+--------------------+--+
| test1            |
| test2            |
+--------------------+--+
```

4．当前数据库

使用 USE <数据库名>设置当前数据库，这样在指定数据表名时，不需要再以<数据库名>.<表名>的形式，而可以直接使用表名。

```
USE test1;
```

使用 SELECT current_database( )查看当前数据库。

```
SELECT current_database( );
```

```
+-----------+--+
|   _c0     |
+-----------+--+
| test1     |
+-----------+--+
```

5. 删除数据库

使用 DROP DATABASE [IF EXISTS] <数据库名> [CASCADE]语句删除数据库 test2，其中：

- IF EXISTS 关键词表示仅在该数据库存在时才删除；
- CASCADE 关键词表示删除数据库的同时也删除数据库中的所有表，如果没有该关键词且数据库中还有表，则会删除失败。

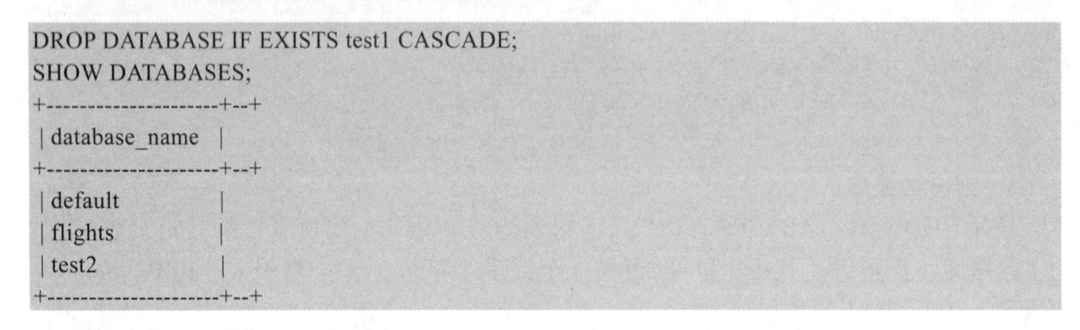

```
DROP DATABASE IF EXISTS test1 CASCADE;
SHOW DATABASES;
+---------------------+--+
| database_name       |
+---------------------+--+
| default             |
| flights             |
| test2               |
+---------------------+--+
```

6.2 数据表基本操作

1. 创建数据表

创建数据表的语句功能强大，形式多样，基本语法和常用选项如下：

```
CREATE [EXTERNAL] TABLE [IF NOT EXISTS] [<数据库名>.]<表名>
  [(<列名 1> <数据类型 1>, <列名 2> <数据类型 2>, …)]
  [PARTITIONED BY (<分区列名 1> <数据类型 1>, <分区列名 2> <数据类型 2>, …)]
  [ROW FORMAT <行格式>]
  [STORED AS <文件存储格式>]
  [LOCATION <HDFS 存储路径>]
  [AS <查询语句>];
```

这些关键词和语句都会在后续章节详细介绍。

使用 CREATE TABLE [IF NOT EXISTS] <表名>语句创建数据表 flights 用于存储航班数据集，其中：

- IF NOT EXISTS 关键词表示仅在该数据表不存在时才创建；
- 列 carrier、tailnum、origin 和 dest 为字符串类型（string），其他列都是整数类型（int），数据类型会在后续章节详细介绍。

```
USE default;
CREATE TABLE IF NOT EXISTS flights
(year int, month int, day int, dep_time int ,dep_delay int, arr_time int, arr_delay int, cancelled int, carrier string,
tailnum string, flight int, origin string, dest string, air_time int, distance int, hour int, min int);
```

2．描述数据表

使用 DESCRIBE [FORMATTED] <表名>语句描述数据表 flights。

```
DESCRIBE flights;
+---------------+---------------+---------------+--+
|   col_name | data_type   | comment   |
+---------------+---------------+---------------+--+
| year        | int          |            |
| month       | int          |            |
| day         | int          |            |
| dep_time    | int          |            |
| dep_delay   | int          |            |
| arr_time    | int          |            |
| arr_delay   | int          |            |
| cancelled   | int          |            |
| carrier     | string       |            |
| tailnum     | string       |            |
| flight      | int          |            |
| origin      | string       |            |
| dest        | string       |            |
| air_time    | int          |            |
| distance    | int          |            |
| hour        | int          |            |
| min         | int          |            |
+---------------+--------------+--------------+--+
```

使用 FORMATTED 关键词以表格形式显示数据表的额外信息。

```
DESCRIBE FORMATTED flights;
+------------------------------+----------------------------------------------------+------------------------------+--+
|        col_name              |                    data_type                       |          comment             |
+------------------------------+----------------------------------------------------+------------------------------+--+
| # col_name                   | data_type                                          | comment                      |
|                              | NULL                                               | NULL                         |
| year                         | int                                                |                              |
| month                        | int                                                |                              |
| day                          | int                                                |                              |
| dep_time                     | int                                                |                              |
| dep_delay                    | int                                                |                              |
| arr_time                     | int                                                |                              |
| arr_delay                    | int                                                |                              |
| cancelled                    | int                                                |                              |
| carrier                      | string                                             |                              |
| tailnum                      | string                                             |                              |
| flight                       | int                                                |                              |
| origin                       | string                                             |                              |
| dest                         | string                                             |                              |
| air_time                     | int                                                |                              |
| distance                     | int                                                |                              |
| hour                         | int                                                |                              |
| min                          | int                                                |                              |
```

```
|                             | NULL                                          | NULL                         |
| # Detailed Table Information   | NULL                                     | NULL                         |
| Database:                   | default                                       | NULL                         |
| Owner:                      | root                                          | NULL                         |
| CreateTime:                 | Tue Oct 02 08:34:32 UTC 2018                  | NULL                         |
| LastAccessTime:             | UNKNOWN                                       | NULL                         |
| Protect Mode:               | None                                          | NULL                         |
| Retention:                  | 0                                             | NULL                         |
| Location:                   | hdfs://master:8020/user/hive/warehouse/flights  | NULL                       |
| Table Type:                 | MANAGED_TABLE                                 | NULL                         |
| Table Parameters:           | NULL                                          | NULL                         |
|                             | transient_lastDdlTime                         | 1538469272                   |
|                             | NULL                                          | NULL                         |
| # Storage Information       | NULL                                          | NULL                         |
| SerDe Library:              | org.apache.hadoop.hive.serde2.lazy.LazySimpleSerDe        | NULL             |
| InputFormat:                | org.apache.hadoop.mapred.TextInputFormat                  | NULL             |
| OutputFormat:               | org.apache.hadoop.hive.ql.io.HiveIgnoreKeyTextOutputFormat | NULL            |
| Compressed:                 | No                                            | NULL                         |
| Num Buckets:                | -1                                            | NULL                         |
| Bucket Columns:             | []                                            | NULL                         |
| Sort Columns:               | []                                            | NULL                         |
| Storage Desc Params:        | NULL                                          | NULL                         |
|                             | serialization.format                          | 1                            |
+-----------------------------+-----------------------------------------------+------------------------------+--+
```

3．列出数据表

使用 SHOW TABLES [LIKE '<正则表达式>']语句列出所有（满足正则表达式的）数据表。正则表达式中，星号*指代一个或多个字母，竖线|表示或。

```
SHOW TABLES;
+---------------+--+
| tab_name   |
+---------------+--+
| flights       |
+---------------+--+
SHOW TABLES LIKE 'f*';
+---------------+--+
| tab_name   |
+---------------+--+
| flights       |
+---------------+--+
```

4．删除数据表

使用 DROP TABLE [IF EXISTS] <表名>语句删除表 flights，其中：IF EXISTS 关键词表示仅在该表存在时才删除。

删除表时会同时删除表的元数据和数据。事实上，数据会移动到 HDFS 上的.Trash/Current 文件夹中。

```
DROP TABLE IF EXISTS flights;
```

5. 从查询结果创建数据表

使用 CREATE TABLE [IF NOT EXISTS] <表名> AS <查询语句>语句从查询结果创建数据表 flights，查询结果为简单选取 flights.flights14 表中的所有数据。

```
CREATE TABLE IF NOT EXISTS flights_top5 AS
SELECT * FROM flights.flights14 limit 5;
```

6. 复制表结构

使用 CREATE TABLE [IF NOT EXISTS] <新表名> LIKE <原表名>语句复制 flights.flights14 表结构创建数据表 flights，并不复制数据。

```
CREATE TABLE IF NOT EXISTS flights LIKE flights.flights14;
```

6.3 存储格式和行格式

创建表时，STORED AS 语句用于指定表中数据的存储格式，常用选项有 TEXTFILE、SEQUENCEFILE、ORC 和 PARQUET 等。其中，TEXTFILE 表示文本文件，也是表中数据的默认存储格式，以可读的明文存储数据，但所占空间较大，性能较差。其他存储格式都是二进制格式，进行了压缩和性能优化。

1. 文本存储格式

在创建表 flights 时，使用 STORED AS TEXFILE 语句表示存储为文本文件，该语句也可以省略，因为文本文件是默认存储格式。

```
DROP TABLE IF EXISTS flights_top5;
CREATE TABLE IF NOT EXISTS flights_top5
STORED AS TEXTFILE AS
SELECT * FROM flights.flights14 limit 5;
```

文本文件都是可读的明文，使用 dfs -cat 命令显示文件内容。

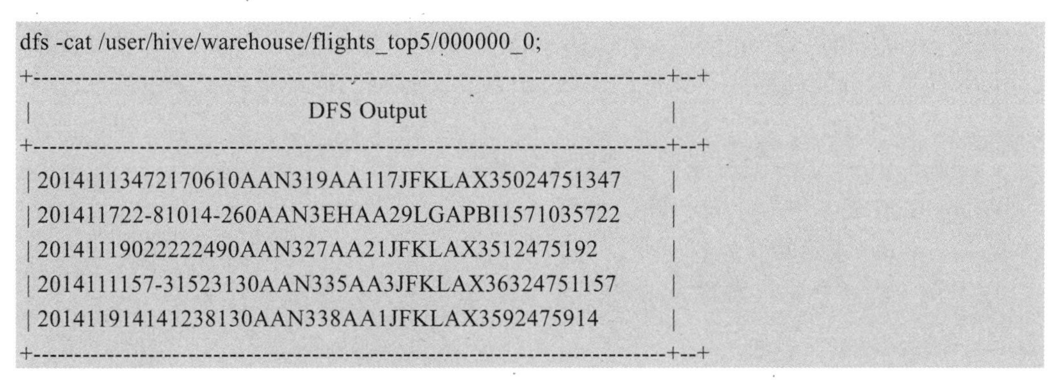

```
dfs -cat /user/hive/warehouse/flights_top5/000000_0;
+------------------------------------------------------------------------------+--+
|                              DFS Output                                      |
+------------------------------------------------------------------------------+--+
| 20141113472170610AAN319AA117JFKLAX35024751347                                |
| 201411722-81014-260AAN3EHAA29LGAPBI1571035722                                |
| 2014111902222490AAN327AA21JFKLAX3512475192                                   |
| 2014111157-31523130AAN335AA3JFKLAX36324751157                                |
| 2014119141412381 30AAN338AA1JFKLAX3592475914                                 |
+------------------------------------------------------------------------------+--+
```

可以看出，每一列的值看似直接连接在了一起。事实上，每一行的格式是以不可见字符'\001'分隔的列。

2. 行格式

对于文本存储格式，行的默认格式是以指定字符分隔的列。

使用 ROW FORMAT 语句设置行格式，这里，DELIMITED 关键词表示行格式是以指定字符分隔的列；FIELDS TERMINATED BY 语句表示分隔符，这里设为逗号。

```
DROP TABLE IF EXISTS flights_top5;
CREATE TABLE IF NOT EXISTS flights_top5
ROW FORMAT DELIMITED FIELDS TERMINATED BY ','
STORED AS TEXTFILE
AS
SELECT * FROM flights.flights14 limit 5;
```

使用 dfs -cat 命令显示文件内容。

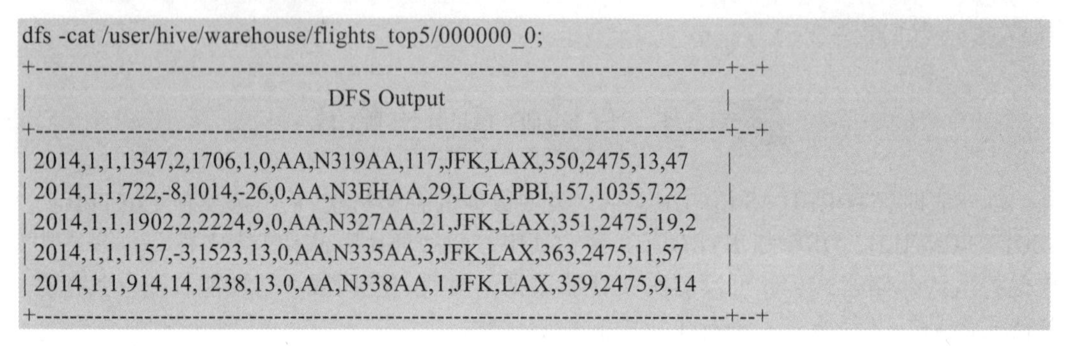

```
dfs -cat /user/hive/warehouse/flights_top5/000000_0;
+--------------------------------------------------------------------------------+--+
|                                  DFS Output                                    |
+--------------------------------------------------------------------------------+--+
| 2014,1,1,1347,2,1706,1,0,AA,N319AA,117,JFK,LAX,350,2475,13,47                  |
| 2014,1,1,722,-8,1014,-26,0,AA,N3EHAA,29,LGA,PBI,157,1035,7,22                  |
| 2014,1,1,1902,2,2224,9,0,AA,N327AA,21,JFK,LAX,351,2475,19,2                    |
| 2014,1,1,1157,-3,1523,13,0,AA,N335AA,3,JFK,LAX,363,2475,11,57                  |
| 2014,1,1,914,14,1238,13,0,AA,N338AA,1,JFK,LAX,359,2475,9,14                    |
+--------------------------------------------------------------------------------+--+
```

可以看出，每一行的格式是以逗号分隔的列。

分隔符为逗号的文本文件又称 CSV（comma-seperated values）文件，分隔符为制表符<Tab>的文件又称 TSV（tab-seperated values）文件。

Hive 还提供多个功能强大的序列化和反序列化器用于指定行格式，具体语法为：

```
ROW FORMAT SERDE '<序列化和反序列化器类名>'
[WITH SERDEPROPERTIES (
    "<属性名 1>" = "<属性值 1>",
    "<属性名 2>" = "<属性值 2>",
    ...)]
```

使用 org.apache.hadoop.hive.serde2.OpenCSVSerde 序列化和反序列化器，设置行格式为以指定字符分隔的列，达到使用 DELIMITED 关键词相同的效果，其中可以进一步设置以下表属性：

- separatorChar 表示列分隔符，默认为逗号；
- quoteChar 表示引号字符，默认为双引号；
- escapeChar 表示转义字符，默认为反斜杠。

以下例子将列分隔符设置为制表符<Tab>，引号字符设置为单引号'。

```
DROP TABLE IF EXISTS flights_top5;
CREATE TABLE IF NOT EXISTS flights_top5
ROW FORMAT SERDE 'org.apache.hadoop.hive.serde2.OpenCSVSerde'
WITH SERDEPROPERTIES (
    "separatorChar" = "\t",
    "quoteChar" = "'"
)
STORED AS TEXTFILE
```

```
AS
SELECT * FROM flights.flights14 limit 5;
```

使用 dfs –cat 命令显示文件内容。

```
dfs -cat /user/hive/warehouse/flights_top5/000000_0;
+------------------------------------------------------------------------------------------------+--+
|                                    DFS Output                                                  |
+------------------------------------------------------------------------------------------------+--+
| '2014' '1' '1' '1347' '2' '1706' '1' '0' 'AA' 'N319AA' '117' 'JFK' 'LAX' '350' '2475' '13' '47'  |
| '2014' '1' '1' '722' '-8' '1014' '-26' '0' 'AA' 'N3EHAA' '29' 'LGA' 'PBI' '157' '1035' '7' '22'  |
| '2014' '1' '1' '1902' '2' '2224' '9' '0' 'AA' 'N327AA' '21' 'JFK' 'LAX' '351' '2475' '19' '2'    |
| '2014' '1' '1' '1157' '-3' '1523' '13' '0' 'AA' 'N335AA' '3' 'JFK' 'LAX' '363' '2475' '11' '57'  |
| '2014' '1' '1' '914' '14' '1238' '13' '0' 'AA' 'N338AA' '1' 'JFK' 'LAX' '359' '2475' '9' '14'    |
+------------------------------------------------------------------------------------------------+--+
```

可以看出，每一行的格式是以制表符<Tab>分隔且包含在单引号内的列。

3. 二进制存储格式

除了文本文件以外的存储格式都称为二进制存储格式，包括 SEQUENCEFILE、ORC 和 PARQUET 等。

使用 STORED AS SEQUENCEFILE 语句表示存储为压缩的序列文件。

```
DROP TABLE IF EXISTS flights_top5;
CREATE TABLE IF NOT EXISTS flights_top5
STORED AS SEQUENCEFILE
AS
SELECT * FROM flights.flights14 limit 5;
```

使用 dfs –cat 命令显示文件内容。

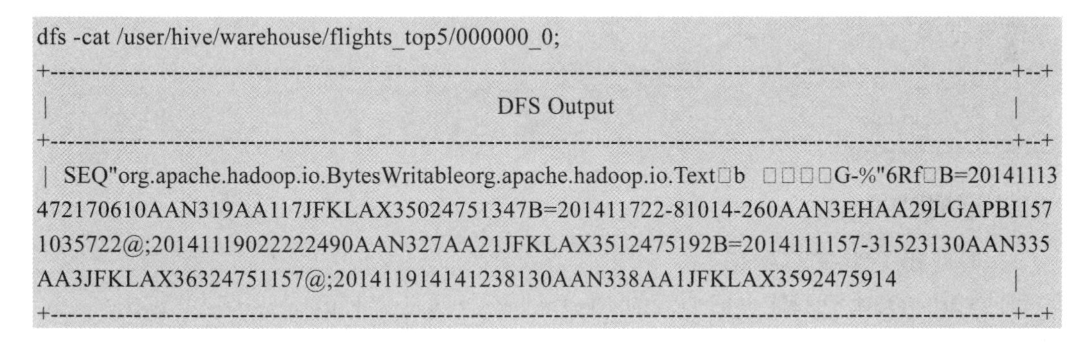

```
dfs -cat /user/hive/warehouse/flights_top5/000000_0;
+------------------------------------------------------------------------------------------------+--+
|                                    DFS Output                                                  |
+------------------------------------------------------------------------------------------------+--+
| SEQ"org.apache.hadoop.io.BytesWritableorg.apache.hadoop.io.Text□b □□□□G-%"6Rf□B=20141113
472170610AAN319AA117JFKLAX35024751347B=201411722-81014-260AAN3EHAA29LGAPBI157
1035722@;20141119022222490AAN327AA21JFKLAX3512475192B=2014111157-31523130AAN335
AA3JFKLAX36324751157@;201411914141238130AAN338AA1JFKLAX3592475914                   |
+------------------------------------------------------------------------------------------------+--+
```

可以看出，文件内容不可读。

6.4 数据类型

Hive 中常用的数据类型主要有以下几种：

1. 数值类型

- 整数（INT）：占用 4 字节，范围从–2 147 483 648 到 2 147 483 647；
- 浮点数（DOUBLE）：占用 8 个字节；

2. 日期时间类型

- 时间戳（TIMESTAMP）；
- 日期（DATE）。

3. 字符串类型。

- 字符串（STRING）。

4. 其他类型

- 布尔（BOOLEAN）。

5. 复杂类型

- 数组（ARRAY<<数据类型>>）；
- 字典（MAP<<键的数据类型>, <值的数据类型>>）；
- 结构体（STRUCT<<列名>：<数据类型>, ...>）。

函数 cast(<列名或数值> as <新数据类型>)用于类型转换。

6.4.1 字符串

输入字符串时，可以使用单引号（'）或双引号（"）。

```
SELECT 'single', "double";
+------------+------------+--+
|    _c0     |    _c1     |
+------------+------------+--+
| single     | double     |
+------------+------------+--+
```

如果输入的字符串中包含引号，则需要通过反斜杠\转义。

```
SELECT '\'single\'', "\"double\"";
+------------+------------+--+
|   _c0      |   _c1      |
+------------+------------+--+
| 'single'   | "double"   |
+------------+------------+--+
```

6.4.2 日期时间

1. 时间戳（TIMESTAMP）

时间戳（TIMESTAMP）的形式为<年（4 位）>-<月（2 位）>-<日（2 位）> <小时（2 位）>:<分钟（2 位）>:<秒（2 位）>。

输入时间戳时，需要在前面加上 timestamp 关键词。

```
SELECT timestamp'2019-01-01 19:18:18';
+------------------------------+--+
|              _c0             |
+------------------------------+--+
| 2019-01-01 19:18:18.0        |
+------------------------------+--+
```

时间戳支持从 UNIX 时间戳的数值型转换得到。

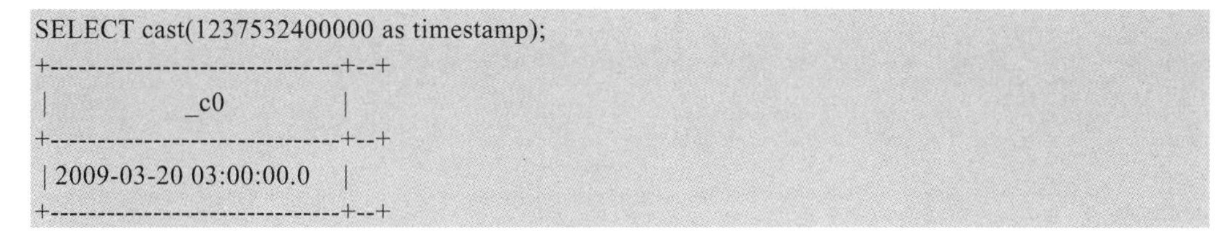

```
SELECT cast(1237532400000 as timestamp);
+------------------------------+--+
|              _c0             |
+------------------------------+--+
| 2009-03-20 03:00:00.0        |
+------------------------------+--+
```

2. 日期（DATE）

日期（DATE）的形式为<年（4 位）>–<月（2 位）>–<日（2 位）>，而不包含时间部分。

输入日期时，需要在前面加上 date 关键词。

```
select date'2019-01-01';
+----------------+--+
|     _c0        |
+----------------+--+
| 2019-01-01     |
+----------------+--+
```

6.4.3 复杂类型

复杂类型主要包括：

- 数组（ARRAY<<数据类型>>）由不定个数且类型相同的数据元素组成；
- 字典（MAP<<键的数据类型>, <值的数据类型>>）由不定个数且类型相同的键值对组成；
- 结构体（STRUCT<<列名 1> : <数据类型 1>, <列名 2> : <数据类型 2>, …>）由个数相同且类型可以不同的数据元素组成。

创建表 complex 包含 3 列，分别为：

- 第 1 列是由字符串组成的数组；
- 第 2 列是由键为整数、值为字符串组成的字典；
- 第 3 列是由 3 个子列组成的结构体，分别为整数 a、浮点数 b 和字符串 c 组成。

对于包含复杂数据类型且行格式是以指定字符分隔的形式的表，还需要进一步在建表语句中指定复杂数据类型中元素的分隔符以及字典中键值对的分隔符。以下例子中：

- 使用 COLLECTION ITEMS TERMINATED BY 语句将复杂数据类型中元素的分隔符设为竖线（|）；
- 使用 MAP KEYS TERMINATED BY 语句将字典中键值对的分隔符设为冒号（:）。

```
CREATE TABLE IF NOT EXISTS complex (
array_col array<string>,
map_col map<int, string>,
struct_col struct<a : int, b : double, c : string>)
ROW FORMAT DELIMITED FIELDS TERMINATED BY ','
COLLECTION ITEMS TERMINATED BY '|'
MAP KEYS TERMINATED BY ':';
DESCRIBE complex;
```

```
+--------------+----------------------------------+------------------+--+
| col_name     |            data_type             |     comment      |
+--------------+----------------------------------+------------------+--+
| array_col    | array<string>                    |                  |
| map_col      | map<int,string>                  |                  |
| struct_col   | struct<a:int,b:double,c:string>  |                  |
+--------------+----------------------------------+------------------+--+
```

复杂类型不支持直接输入，这里创建一个文本文件，并将其导入新创建的表，导入的语法会在后续章节具体介绍。

新启动一个终端，使用 echo 命令写入若干行测试数据到文件 complex.csv。

```
echo "1a|1b|1c|1d,101:1a|102:1b,1|1.0|1a
2a|2b,201:2a|202:2b|203:2c,2|2.0|2a
3a,301:3a,3|3.0|3a" > complex.csv
```

回到 beeline 所在的终端，使用 LOAD DATA 语句导入该文件。

```
LOAD DATA LOCAL INPATH '/root/complex.csv'
OVERWRITE INTO TABLE complex;
SELECT * FROM complex;
+----------------------------+----------------------------------+-----------------------------+--+
|   complex.array_col        |         complex.map_col          |     complex.struct_col      |
+----------------------------+----------------------------------+-----------------------------+--+
| ["1a","1b","1c","1d"]      | {101:"1a",102:"1b"}              | {"a":1,"b":1.0,"c":"1a"}    |
| ["2a","2b"]                | {201:"2a",202:"2b",203:"2c"}     | {"a":2,"b":2.0,"c":"2a"}    |
| ["3a"]                     | {301:"3a"}                       | {"a":3,"b":3.0,"c":"3a"}    |
+----------------------------+----------------------------------+-----------------------------+--+
```

6.5 外 部 表

Hive 默认创建的都是内部表，即表的数据文件、元数据和统计信息都由 Hive 直接管理。Hive 的数据文件默认存储在 HDFS 的/user/hive/warehouse/<数据库名>.db/<表名>/文件夹，可以通过 Hive 配置文件中的 hive.metastore.warehouse.dir 配置项。如果在 Hive 中删除内部表，则与表关联的数据文件和元数据都会被删除，即 Hive 管理表的整个生命周期。

Hive 的外部表仅管理表的元数据，而并不负责管理数据文件。如果在 Hive 中删除外部表，则仅会删除元数据，而不会删除数据文件。

使用 EXTERNAL 关键字创建外部表 flights，并使用 LOCATION 语句指定表的数据文件路径为/user/root/flights14 文件夹。

```
CREATE EXTERNAL TABLE IF NOT EXISTS flights
(year int, month int, day int, dep_time int ,dep_delay int,
arr_time int, arr_delay int, cancelled int, carrier string,
tailnum string, flight int, origin string, dest string,
air_time int, distance int, hour int, min int)
LOCATION '/user/root/flights14';
```

此时由于表 flights 对应的数据文件路径并没有任何数据，因此无法选取出任何数据。

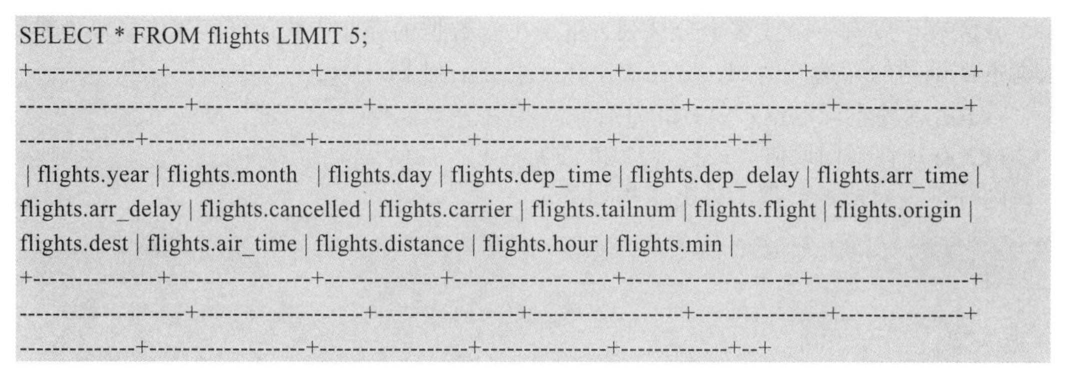

```
SELECT * FROM flights LIMIT 5;
+---------------+----------------+-------------+------------------+--------------------+------------------+
-------------------+-------------------+----------------+-----------------+---------------+--------------+
-------------+------------------+-----------------+--------------+------------+--+
| flights.year | flights.month  | flights.day | flights.dep_time | flights.dep_delay | flights.arr_time |
flights.arr_delay | flights.cancelled | flights.carrier | flights.tailnum | flights.flight | flights.origin |
flights.dest | flights.air_time | flights.distance | flights.hour | flights.min |
+---------------+----------------+-------------+------------------+--------------------+------------------+
-------------------+-------------------+----------------+-----------------+---------------+--------------+
-------------+------------------+-----------------+--------------+------------+--+
```

使用 dfs –cp 命令将数据文件复制到外部表 flights 对应的数据文件路径。

```
dfs -cp /user/hive/warehouse/flights.db/flights14/ /user/root/;
```

此时则可以查询到数据。

```
SELECT * FROM flights LIMIT 5;
+---------------+----------------+-------------+------------------+--------------------+------------------+
-------------------+-------------------+----------------+-----------------+---------------+--------------+
-------------+------------------+-----------------+--------------+------------+--+
| flights.year | flights.month  | flights.day | flights.dep_time | flights.dep_delay | flights.arr_time |
flights.arr_delay | flights.cancelled | flights.carrier | flights.tailnum | flights.flight | flights.origin |
flights.dest | flights.air_time | flights.distance | flights.hour | flights.min |
+-----+-----+-----+-----+-----+-----+-----+-----+-----+-----------+-----+-----+-----+-----+-----+-----+-----+--+
|2014|1    |1    |914  |14   |1238|13   |0    |AA   |N338AA    |1    |JFK  |LAX|359  |2475|9    |14   |
|2014|1    |1    |1157|-3   |1523|13   |0    |AA   |N335AA    |3    |JFK  |LAX|363  |2475|11   |57   |
|2014|1    |1    |1902|2    |2224|9    |0    |AA   |N327AA    |21   |JFK  |LAX|351  |2475|19   |2    |
|2014|1    |1    |722  |-8   |1014|-26  |0    |AA   |N3EHAA    |29   |LGA|PBI  |157  |1035|7    |22   |
|2014|1    |1    |1347|2    |1706|1    |0    |AA   |N319AA    |117 |JFK  |LAX|350  |2475|13   |47   |
+-----+-----+-----+-----+-----+-----+-----+-----+-----+-----------+-----+-----+-----+-----+-----+-----+-----+--+
```

使用 DROP TABLE 语句删除外部表 flights。

```
DROP TABLE flights;
```

使用 dfs –ls 查看原来的表 flights 对应的数据文件路径/user/root/flights14。

```
dfs -ls /user/root/flights14;
+----------------------------------------------------------------------------------------------------+--+
|                                              DFS Output                                            |
+----------------------------------------------------------------------------------------------------+--+
| Found 1 items                                                                                      |
| -rw-r--r--    1 root hadoop     16150340 2018-10-03 02:43 /user/root/flights14/000000_0 |
+----------------------------------------------------------------------------------------------------+--+
```

可以看出，数据文件仍然存在，并没有因为外部表的删除而被删除。

6.6 分 区 表

分区表可以有一个或多个分区列，用于决定数据的存储方式。分区还能更高效地依据条件做行的筛选。例如，分区表 t 有 2 个分区列 k1 和 k2，其中：

- k1 有 2 种不同取值：v11 和 v12；
- k2 有 3 种不同取值：v21、v22 和 v23。

则分区表对应的数据文件夹结构如下：

```
't'
--'k1=v11'
----'k2=v21'
----'k2=v22'
----'k2=v23'
--'k1=v12'
----'k2=v21'
----'k2=v22'
----'k2=v23'
```

当查询条件中包含分区列时，如仅需要查询 k1=v11 和 k2=v23 时，则可以直接定位到相应的数据文件夹 t/k1=v11/k2=v23，因此更高效。

使用 PARTITIONED BY (<分区列名 1> <数据类型 1>, <分区列名 2> <数据类型 2>, …)语句指定表 flights 的分区，这里的分区列为 year 和 month。

```
DROP TABLE IF EXISTS flights;
CREATE TABLE IF NOT EXISTS flights
(day int, dep_time int ,dep_delay int,
arr_time int, arr_delay int, cancelled int, carrier string,
tailnum string, flight int, origin string, dest string,
air_time int, distance int, hour int, min int)
PARTITIONED BY (year int, month int);
```

使用 DESCRIBE 语句描述表 flights。

```
DESCRIBE flights;
+------------------------------+-----------------------+-----------------------+--+
|           col_name           |       data_type       |        comment        |
+------------------------------+-----------------------+-----------------------+--+
| day                          | int                   |                       |
| dep_time                     | int                   |                       |
| dep_delay                    | int                   |                       |
| arr_time                     | int                   |                       |
| arr_delay                    | int                   |                       |
| cancelled                    | int                   |                       |
| carrier                      | string                |                       |
| tailnum                      | string                |                       |
| flight                       | int                   |                       |
| origin                       | string                |                       |
| dest                         | string                |                       |
| air_time                     | int                   |                       |
```

```
| distance                    | int                   |                       |
| hour                        | int                   |                       |
| min                         | int                   |                       |
| year                        | int                   |                       |
| month                       | int                   |                       |
|                             | NULL                  | NULL                  |
| # Partition Information     | NULL                  | NULL                  |
| # col_name                  | data_type             | comment               |
|                             | NULL                  | NULL                  |
| year                        | int                   |                       |
| month                       | int                   |                       |
+-----------------------------+-----------------------+-----------------------+--+
```

可以看出，最下面的 Partition Information 区域显示了表的分区列。

使用 2 条 INSERT 语句插入 2014 年 1 月和 2 月的数据到分区表 flights 中，插入时需要指定分区。插入数据的语法会在后续章节具体介绍。

```
INSERT OVERWRITE TABLE flights PARTITION (year=2014, month=1)
SELECT day, dep_time, dep_delay, arr_time, arr_delay, cancelled, carrier,
tailnum, flight, origin, dest, air_time, distance, hour, min
FROM flights.flights14
WHERE year=2014 and month=1;

INSERT OVERWRITE TABLE flights PARTITION (year=2014, month=2)
SELECT day, dep_time, dep_delay, arr_time, arr_delay, cancelled, carrier,
tailnum, flight, origin, dest, air_time, distance, hour, min
FROM flights.flights14
WHERE year=2014 and month=2;
```

使用 dfs –ls 查看分区表 flights 对应的数据文件路径/user/hive/warehouse/flights 的具体结构。

```
dfs -ls /user/hive/warehouse/flights/*;
+------------------------------------------------------------------------------------------------+--+
|                                    DFS Output                                                  |
+------------------------------------------------------------------------------------------------+--+
| Found 2 items                                                                                  |
| drwxrwxrwt - root hadoop 0 2018-10-03 05:59 /user/hive/warehouse/flights/year=2014/month=1 |
| drwxrwxrwt - root hadoop 0 2018-10-03 05:59 /user/hive/warehouse/flights/year=2014/month=2 |
+------------------------------------------------------------------------------------------------+--+
```

使用 SHOW PARTITIONS 语句显示表 flights 的分区。

```
SHOW PARTITIONS flights;
+---------------------------+--+
|         partition         |
+---------------------------+--+
| year=2014/month=1         |
| year=2014/month=2         |
+---------------------------+--+
```

小　结

本章主要介绍了 Hive 数据仓库的数据定义操作，首先介绍了数据库、数据表、分区和桶或簇的概念。然后介绍了数据库操作和数据表操作的具体语法，包括创建、描述、列出、删除等操作。接着介绍了存储格式和数据类型。Hive 数据表可以存储为文本文件，也可以存储为二进制格式，进行了压缩和性能优化。

本章还着重介绍了 Hive 数据仓库相对传统数据库而言的独有特性，即外部表和分区表。Hive 的外部表仅管理表的元数据，而并不负责管理数据文件。分区表可以有一个或多个分区列，用于决定数据的存储方式。分区还能更高效地依据条件做行的筛选。

习　题

1. Hive 从粗到细、从顶层到底层的组织方式是什么？
2. 在 Hive 中执行数据库的创建、描述、列出、删除等操作。
3. 在 Hive 中执行数据表的创建、描述、列出、删除等操作。
4. 在 Hive 中创建数组、字典和结构体等复杂数据类型。
5. 在 Hive 中创建外部表，并删除，观察数据文件是否存在。
6. 在 Hive 中创建分区表，观察数据文件夹结构。

Hive 数据操作 ‹‹‹

本章所有例子同样基于航班的准点情况数据，之前章节已有介绍。该数据集的路径为～/data/flights/flights14.csv，其中第一行为列名。同时也已经存储在 Hive 的 flights.flights14 表中。

7.1 数据导入

数据导入语句的基本语法和常用选项如下：

```
LOAD DATA [LOCAL] INPATH '<数据文件（夹）路径>'
[OVERWRITE] INTO TABLE <表名> [PARTITION (<分区列名 1>=<分区列值 1>, <分区列名 2>=
<分区列值 2>, ...)]
```

该语句仅仅是将数据文件复制或移动到 Hive 表对应的文件夹下。

1. 从本地文件系统导入数据表

使用 CREATE TABLE 语句创建数据表 flights 用于存储航班数据集。

```
CREATE TABLE IF NOT EXISTS flights
(year int, month int, day int, dep_time int ,dep_delay int,
arr_time int, arr_delay int, cancelled int, carrier string,
tailnum string, flight int, origin string, dest string,
air_time int, distance int, hour int, min int)
ROW FORMAT DELIMITED FIELDS TERMINATED BY ','
STORED AS TEXTFILE;
```

使用 LOAD DATA LOCAL INPATH '<数据文件路径>' INTO TABLE <表名>语句将本地文件系统中的数据文件/root/data/flights/flights14.csv 导入表 flights。其中，LOCAL 关键字表示从本地文件系统导入。

```
LOAD DATA LOCAL INPATH '/root/data/flights/flights14.csv'
INTO TABLE flights;
[...]
INFO   : Starting task [Stage-0:MOVE] in serial mode
INFO   : Loading data to table default.flights from file:/root/data/flights/flights14.csv
INFO   : Starting task [Stage-1:STATS] in serial mode
[...]
SELECT * FROM flights LIMIT 5;
```

```
+---------------+-----------------+--------------+-------------------+---------------------+-------------------+
--------------------+--------------------+-----------------+------------------+----------------+---------------+
--------------+-------------------+------------------+---------------+-------------+--+
| flights.year | flights.month  | flights.day | flights.dep_time | flights.dep_delay | flights.arr_time |
flights.arr_delay | flights.cancelled | flights.carrier | flights.tailnum | flights.flight | flights.origin |
flights.dest | flights.air_time | flights.distance | flights.hour | flights.min |
+-----+-----+-----+-----+-----+-----+-----+-----+-----+-----------+-----+-----+-----+-----+-----+-----+-----+--+
|NULL |NULL |NULL |NULL |NULL |NULL |NULL |NULL |carrier |tailnum     |NULL |origin |dest |NULL |NULL |NULL |NULL |
|2014|1     |1     |914  |14   |1238|13    |0     |AA    |N338AA  |1    |JFK |LAX|359  |2475|9    |14   |
|2014|1     |1     |1157|-3    |1523|13    |0     |AA    |N335AA  |3    |JFK |LAX|363  |2475|11   |57   |
|2014|1     |1     |1902|2     |2224|9     |0     |AA    |N327AA  |21   |JFK |LAX|351  |2475|19   |2    |
|2014|1     |1     |722  |-8    |1014|-26   |0     |AA    |N3EHAA  |29   |LGA|PBI |157  |1035|7    |22   |
+-----+-----+-----+-----+-----+-----+-----+-----+-----+-----------+-----+-----+-----+-----+-----+-----+-----+--+
```

第一行的所有列都是缺失值，因为数据文件的第一行是列名。

该语句仅仅是将数据文件从本地文件系统上传到 Hive 表对应的路径下。

2．从 HDFS 导入数据表

重新创建数据表 flights。

```
DROP TABLE IF EXISTS flights;
CREATE TABLE IF NOT EXISTS flights
(year int, month int, day int, dep_time int ,dep_delay int,
arr_time int, arr_delay int, cancelled int, carrier string,
tailnum string, flight int, origin string, dest string,
air_time int, distance int, hour int, min int)
ROW FORMAT DELIMITED FIELDS TERMINATED BY ','
STORED AS TEXTFILE;
```

新启动一个终端，使用 hdfs dfs –put 命令将数据文件 flights14 上传到 HDFS。

```
hdfs dfs -put data/flights/flights14.csv
```

使用 LOAD DATA INPATH '<数据文件路径>' INTO TABLE <表名>语句将 HDFS 中的数据文件/user/root/flights14.csv 导入表 flights。

```
LOAD DATA INPATH '/user/root/flights14.csv' INTO TABLE flights;
[...]
INFO    : Starting task [Stage-0:MOVE] in serial mode
INFO    : Loading data to table default.flights from hdfs://master:8020/user/root/flights14.csv
INFO    : Starting task [Stage-1:STATS] in serial mode
[...]
```

该语句仅仅是将数据文件在 HDFS 上从原路径**移动**到 Hive 表对应的路径下。

```
SELECT * FROM flights limit 5;
+---------------+-----------------+--------------+-------------------+---------------------+-------------------+
--------------------+--------------------+-----------------+------------------+----------------+---------------+
--------------+-------------------+------------------+---------------+-------------+--+
```

```
| flights.year | flights.month  | flights.day | flights.dep_time | flights.dep_delay | flights.arr_time |
flights.arr_delay | flights.cancelled | flights.carrier | flights.tailnum | flights.flight | flights.origin |
flights.dest | flights.air_time | flights.distance | flights.hour | flights.min |
+-----+-----+-----+-----+-----+-----+-----+-----+-----+-----------+-----+-----+-----+-----+-----+-----+-----+--+
| NULL | NULL | NULL | NULL | NULL | NULL | NULL | NULL | carrier | tailnum | NULL | origin | dest | NULL | NULL | NULL | NULL |
| 2014| 1   | 1   | 914  | 14  | 1238| 13  | 0   | AA  | N338AA  | 1   | JFK | LAX| 359 | 2475| 9   | 14  |
| 2014| 1   | 1   | 1157| -3  | 1523| 13  | 0   | AA  | N335AA  | 3   | JFK | LAX| 363 | 2475| 11  | 57  |
| 2014| 1   | 1   | 1902| 2   | 2224| 9   | 0   | AA  | N327AA  | 21  | JFK | LAX| 351 | 2475| 19  | 2   |
| 2014| 1   | 1   | 722  | -8  | 1014| -26 | 0   | AA  | N3EHAA  | 29  | LGA| PBI | 157 | 1035| 7   | 22  |
+-----+-----+-----+-----+-----+-----+-----+-----+-----+-----------+-----+-----+-----+-----+-----+-----+-----+--+
```

使用 dfs –ls 命令原数据文件统计信息。

```
dfs -ls /user/root;
+------------------+
| DFS Output   |
+------------------+
+------------------+
```

可以看出，原数据文件已经不存在，因为已经移动到 Hive 表对应的路径下。

3. 导入并覆盖

由于表 flights 中已经有数据，如果不使用 OVERWRITE 关键字则会在原有数据基础上添加新导入数据。

```
LOAD DATA LOCAL INPATH '/root/data/flights/flights14.csv' INTO TABLE flights;
[...]
INFO   : Starting task [Stage-0:MOVE] in serial mode
INFO   : Loading data to table default.flights from file:/root/data/flights/flights14.csv
INFO   : Starting task [Stage-1:STATS] in serial mode
[...]
```

使用 dfs –ls 命令显示表 flights 对应文件夹的统计信息。

```
dfs -ls /user/hive/warehouse/flights;
+----------------------------------------------------------------------------------------------------------------+
|                                               DFS Output                                                        |
+----------------------------------------------------------------------------------------------------------------+
| Found 2 items                                                                                                   |
| -rwxrwxrwt 1 root hadoop 16150465 2019-02-05 05:57 /user/hive/warehouse/flights/flights14.csv         |
| -rwxrwxrwt 1 root hadoop 16150465 2019-02-05 06:01 /user/hive/warehouse/flights/flights14_copy_1.csv |
+----------------------------------------------------------------------------------------------------------------+
```

可以看出，该表对应的数据文件有 2 个，且文件大小完全一样。

使用 OVERWRITE 关键字导入并覆盖，将仅保留新导入数据。

```
LOAD DATA LOCAL INPATH '/root/data/flights/flights14.csv' OVERWRITE INTO TABLE flights;
[...]
INFO   : Starting task [Stage-0:MOVE] in serial mode
```

```
INFO   : Loading data to table default.flights from file:/root/data/flights/flights14.csv
INFO   : Starting task [Stage-1:STATS] in serial mode
[...]
```

使用 dfs –ls 命令显示表 flights 对应文件夹的统计信息。

```
dfs -ls /user/hive/warehouse/flights;
+----------------------------------------------------------------------------------------------------+
|                                    DFS Output                                                      |
+----------------------------------------------------------------------------------------------------+
| Found 1 items                                                                                      |
| -rwxrwxrwt 1 root hadoop 16150465 2019-02-05 06:04 /user/hive/warehouse/flights/flights14.csv |
+----------------------------------------------------------------------------------------------------+
```

可以看出，该表对应的数据文件仅有 1 个。

4．导入分区表

使用 CREATE TABLE … PARTITIONED BY …语句创建分区表 flights，其中分区列为 load_date。

```
DROP TABLE IF EXISTS flights;
CREATE TABLE IF NOT EXISTS flights
(day int, dep_time int ,dep_delay int,
arr_time int, arr_delay int, cancelled int, carrier string,
tailnum string, flight int, origin string, dest string,
air_time int, distance int, hour int, min int)
PARTITIONED BY (load_date string);
```

数据导入时，使用 PARTITION 关键词指定导入的分区。

```
LOAD DATA LOCAL INPATH '/root/data/flights/flights14.csv' INTO TABLE flights PARTITION
(load_date='2018-10-12');
[...]
INFO : Starting task [Stage-0:MOVE] in serial mode
INFO : Loading data to table default.flights partition (load_date=2018-10-12) from file:/root/data/flights/
flights14.csv
INFO : Starting task [Stage-1:STATS] in serial mode
[...]
```

7.2 数据插入

数据插入语句的基本语法和常用选项如下：

```
INSERT INTO|OVERWRITE TABLE <表名> [PARTITION (<分区列名 1>=<分区列值 1>, <分区列名
2>=<分区列值 2>, ...)] <查询语句>;
```

1．基本插入

使用 CREATE TABLE 语句创建数据表 flights 用于存储航班数据集。

```
DROP TABLE IF EXISTS flights;
CREATE TABLE IF NOT EXISTS flights
```

```
(year int, month int, day int, dep_time int ,dep_delay int,
arr_time int, arr_delay int, cancelled int, carrier string,
tailnum string, flight int, origin string, dest string,
air_time int, distance int, hour int, min int);
```

使用 INSERT INTO TABLE <表名> <查询语句>语句将查询结果插入表 flights，查询结果为简单选取 flights.flights14 表中的所有数据。

```
INSERT INTO TABLE flights
SELECT * FROM flights.flights14;
[...]
INFO   : Hadoop job information for Stage-1: number of mappers: 1; number of reducers: 0
INFO   : 2018-10-13 10:12:53,595 Stage-1 map = 0%,   reduce = 0%
INFO   : 2018-10-13 10:13:04,228 Stage-1 map = 100%,   reduce = 0%, Cumulative CPU 3.74 sec
[...]
INFO   : Starting task [Stage-4:MOVE] in serial mode
INFO   : Moving data to directory hdfs://master:8020/user/hive/warehouse/flights/.hive-staging_hive_2018-10-13_10-12-44_225_3245717840633687970-3/-ext-10000 from hdfs://master:8020/user/hive/warehouse/flights/.hive-staging_hive_2018-10-13_10-12-44_225_3245717840633687970-3/-ext-10002
INFO   : Starting task [Stage-0:MOVE] in serial mode
INFO   : Loading data to table default.flights from hdfs://master:8020/user/hive/warehouse/flights/.hive-staging_hive_2018-10-13_10-12-44_225_3245717840633687970-3/-ext-10000
INFO   : Starting task [Stage-2:STATS] in serial mode
INFO   : MapReduce Jobs Launched:
INFO   : Stage-Stage-1: Map: 1 Cumulative CPU: 3.3 sec HDFS Read: 16156866 HDFS Write: 16150421 HDFS EC Read: 0 SUCCESS
```

该语句并不会覆盖表中原有数据，而会在原有数据基础上插入新的查询结果。

2. 插入并覆盖

由于表 flights 中已经有数据，如果不使用 OVERWRITE 关键字则会将新导入数据与原有数据按行合并。

```
INSERT INTO TABLE flights
SELECT * FROM flights.flights14;
[...]
INFO   : Hadoop job information for Stage-1: number of mappers: 1; number of reducers: 0
INFO   : 2018-10-13 10:14:56,707 Stage-1 map = 0%,   reduce = 0%
INFO   : 2018-10-13 10:15:03,192 Stage-1 map = 100%,   reduce = 0%, Cumulative CPU 3.76 sec
[...]
INFO   : Starting task [Stage-4:MOVE] in serial mode
INFO   : Moving data to directory hdfs://master:8020/user/hive/warehouse/flights/.hive-staging_hive_2018-10-13_10-14-48_107_2508054682749578220-3/-ext-10000 from hdfs://master:8020/user/hive/warehouse/flights/.hive-staging_hive_2018-10-13_10-14-48_107_2508054682749578220-3/-ext-10002
INFO   : Starting task [Stage-0:MOVE] in serial mode
INFO   : Loading data to table default.flights from hdfs://master:8020/user/hive/warehouse/flights/.hive-staging_hive_2018-10-13_10-14-48_107_2508054682749578220-3/-ext-10000
INFO   : Starting task [Stage-2:STATS] in serial mode
INFO   : MapReduce Jobs Launched:
```

```
INFO   : Stage-Stage-1: Map: 1 Cumulative CPU: 3.29 sec HDFS Read: 16156866 HDFS Write: 16150421
HDFS EC Read: 0 SUCCESS
[...]
```

使用 dfs –ls 命令显示文件夹统计信息。

```
dfs -ls /user/hive/warehouse/flights/;
+----------------------------------------------------------------------------------------------------+
|                                              DFS Output                                            |
+----------------------------------------------------------------------------------------------------+
| Found 2 items                                                                                      |
| -rwxrwxrwt   1 root hadoop     16150340 2018-10-13 10:13 /user/hive/warehouse/flights/000000_0     |
| -rwxrwxrwt   1 root hadoop     16150340 2018-10-13 10:15 /user/hive/warehouse/flights/000000_0_copy_1 |
+----------------------------------------------------------------------------------------------------+
```

可以看出，该表对应的数据文件有 2 个。

使用 INSERT OVERWRITE 语句插入并覆盖，将仅保留新插入数据。

```
INSERT OVERWRITE TABLE flights
SELECT * FROM flights.flights14;
[...]
INFO   : Hadoop job information for Stage-1: number of mappers: 1; number of reducers: 0
INFO   : 2018-10-13 10:19:56,828 Stage-1 map = 0%,   reduce = 0%
INFO   : 2018-10-13 10:20:04,256 Stage-1 map = 100%,   reduce = 0%, Cumulative CPU 4.02 sec
[...]
INFO   : Starting task [Stage-4:MOVE] in serial mode
INFO   : Moving data to directory hdfs://master:8020/user/hive/ warehouse/flights/.hive-staging_hive_
2018-10-13_10-19-47_412_654106681393205683-3/-ext-10000 from hdfs://master:8020/user/hive/
warehouse/flights/.hive-staging_hive_2018-10-13_10-19-47_412_654106681393205683-3/-ext-10002
INFO   : Starting task [Stage-0:MOVE] in serial mode
INFO   : Loading data to table default.flights from hdfs://master:8020/ user/hive/warehouse/flights/.hive-
staging_hive_2018-10-13_10-19-47_412_654106681393205683-3/-ext-10000
INFO   : Starting task [Stage-2:STATS] in serial mode
INFO   : MapReduce Jobs Launched:
INFO   : Stage-Stage-1: Map: 1    Cumulative CPU: 3.26 sec    HDFS Read: 16156786 HDFS Write:
16150421 HDFS EC Read: 0 SUCCESS
[...]
```

使用 dfs –ls 命令显示文件夹统计信息。

```
dfs -ls /user/hive/warehouse/flights/;
+----------------------------------------------------------------------------------------------------+
|                                              DFS Output                                            |
+----------------------------------------------------------------------------------------------------+
| Found 2 items                                                                                      |
| -rwxrwxrwt   1 root hadoop     16150340 2018-10-13 10:20 /user/hive/warehouse/flights/000000_0     |
+----------------------------------------------------------------------------------------------------+
```

可以看出，该表对应的数据文件仅有 1 个。

3．插入分区表

使用 CREATE TABLE … PARTITIONED BY …语句创建分区表 flights，其中分区列为 year 和 month。

```
DROP TABLE IF EXISTS flights;
CREATE TABLE IF NOT EXISTS flights
(day int, dep_time int ,dep_delay int,
arr_time int, arr_delay int, cancelled int, carrier string,
tailnum string, flight int, origin string, dest string,
air_time int, distance int, hour int, min int)
PARTITIONED BY (year int, month int);
```

数据插入时，使用 PARTITION 关键字指定导入的分区。

```
INSERT OVERWRITE TABLE flights PARTITION (year=2014, month=1)
SELECT day, dep_time, dep_delay, arr_time, arr_delay, cancelled, carrier,
tailnum, flight, origin, dest, air_time, distance, hour, min
FROM flights.flights14
WHERE year=2014 and month=1;

INSERT OVERWRITE TABLE flights PARTITION (year=2014, month=2)
SELECT day, dep_time, dep_delay, arr_time, arr_delay, cancelled, carrier,
tailnum, flight, origin, dest, air_time, distance, hour, min
FROM flights.flights14
WHERE year=2014 and month=2;
```

使用 SHOW PARTITIONS 语句显示分区表 flights 的分区。

```
SHOW PARTITIONS flights;
+---------------------------+--+
|         partition         |
+---------------------------+--+
| year=2014/month=1    |
| year=2014/month=2    |
+---------------------------+--+
```

4．动态插入分区表

以上的例子插入分区表时，插入每个分区都需要指定分区列的值，如果分区较多则会非常烦琐。动态分区只需要使用一条语句，自动根据分区列的值，插入相应的分区。

在动态插入分区表前，需要先设置 2 个选项：

- hive.exec.dynamic.partition.mode：表示动态分区模式，strict 模式中用户必须至少指定一个分区以防止用户意外的覆盖所有分区，nonstrict 模式中用户可以不指定任何一个分区，即所有分区都是动态的，默认为 strict；
- hive.exec.dynamic.partition：表示是否允许动态分区，true 表示允许，false 表示不允许，默认为 true。

```
SET hive.exec.dynamic.partition.mode = nonstrict;
SET hive.exec.dynamic.partition = true;
```

使用 INSERT … PARTITION …语句插入分区表。

```
INSERT INTO TABLE flights PARTITION(year, month)
SELECT day, dep_time, dep_delay, arr_time, arr_delay, cancelled, carrier,
tailnum, flight, origin, dest, air_time, distance, hour, min,
```

```
year, month
FROM flights.flights14;
```

使用 SHOW PARTITIONS 语句显示分区表 flights 的分区。

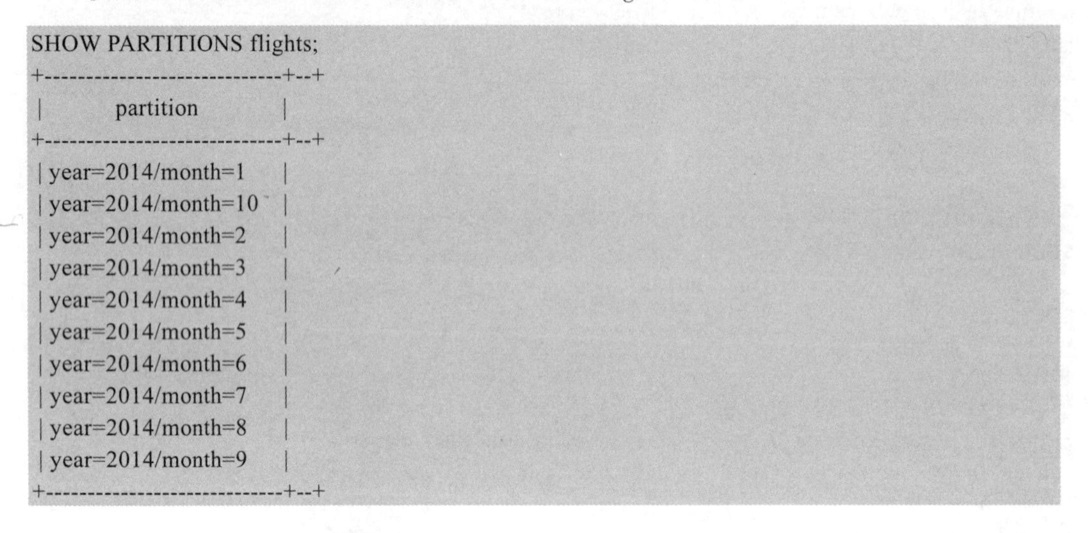

```
SHOW PARTITIONS flights;
+----------------------------+--+
|         partition          |
+----------------------------+--+
| year=2014/month=1    |
| year=2014/month=10   |
| year=2014/month=2    |
| year=2014/month=3    |
| year=2014/month=4    |
| year=2014/month=5    |
| year=2014/month=6    |
| year=2014/month=7    |
| year=2014/month=8    |
| year=2014/month=9    |
+----------------------------+--+
```

7.3 数据导出

数据导出语句的基本语法和常用选项如下：

```
INSERT OVERWRITE [LOCAL] DIRECTORY '<导出路径>'
[ROW FORMAT <行格式>] [STORED AS <文件存储格式>]
<查询语句>
```

1. 导出到本地文件系统

使用 set 命令定义变量 hive.exec.stagingdir 表示 Hive 运行时产生的临时文件存放的路径。

使用 INSERT OVERWRITE LOCAL DIRECTORY '<导出路径>'语句将 flights.flights14 表导出到本地文件系统中的数据文件路径/root/tmp/flights。其中，LOCAL 关键字表示导出到本地文件系统。

```
SET hive.exec.stagingdir=/tmp/;
INSERT OVERWRITE LOCAL DIRECTORY '/root/tmp/flights'
SELECT * FROM flights.flights14;
[...]
INFO  : Hadoop job information for Stage-1: number of mappers: 1; number of reducers: 0
INFO  : 2019-01-22 14:31:16,885 Stage-1 map = 0%,  reduce = 0%
INFO  : 2019-01-22 14:31:26,703 Stage-1 map = 100%,  reduce = 0%, Cumulative CPU 3.31 sec
[...]
INFO  : Starting task [Stage-0:MOVE] in serial mode
INFO  : Moving data to local directory /root/tmp/flights from hdfs://master:8020/tmp/hive/root/a1cbfd7b-defc-47ea-9be3-94c4809ba9fd/hive_2019-01-22_14-31-00_038_2595090853702381322-1/-mr-10000
INFO  : MapReduce Jobs Launched:
```

```
INFO   : Stage-Stage-1: Map: 1 Cumulative CPU: 3.31 sec HDFS Read: 16156187 HDFS Write: 16150340
HDFS EC Read: 0 SUCCESS
[...]
```

使用 head 命令查看导出数据文件的前几行。

```
head /root/tmp/flights/*
201411914141238130AAN338AA1JFKLAX3592475914
2014111157-31523130AAN335AA3JFKLAX36324751157
20141119022222490AAN327AA21JFKLAX3512475192
201411722-81014-260AAN3EHAA29LGAPBI1571035722
20141113472170610AAN319AA117JFKLAX35024751347
20141118244214500AAN3DEAA119EWRLAX33924541824
2014112133-237-180AAN323AA185JFKLAX33824752133
2014111542-31906-140AAN328AA133JFKLAX35624751542
2014111509-11828-170AAN5FJAA145JFKMIA1611089159
2014111848-22206-140AAN3HYAA235JFKSEA34924221848
```

可以看出，每一行的格式是以不可见字符'\001'分隔的列。

数据导出时使用 ROW FORMAT 将示分隔符设为逗号（,）。

```
INSERT OVERWRITE LOCAL DIRECTORY '/root/tmp/flights'
ROW FORMAT DELIMITED FIELDS TERMINATED BY ','
SELECT * FROM flights.flights14;
[...]
INFO   : Hadoop job information for Stage-1: number of mappers: 1; number of reducers: 0
INFO   : 2019-01-22 14:36:37,937 Stage-1 map = 0%,  reduce = 0%
INFO   : 2019-01-22 14:36:47,475 Stage-1 map = 100%,  reduce = 0%, Cumulative CPU 3.09 sec
[...]
INFO   : Starting task [Stage-0:MOVE] in serial mode
INFO   : Moving data to local directory /root/tmp/flights from hdfs://master:8020/tmp/hive/root/a1cbfd7b-
defc-47ea-9be3-94c4809ba9fd/hive_2019-01-22_14-36-23_815_888195784237799922-1/-mr-10000
INFO   : MapReduce Jobs Launched:
INFO   : Stage-Stage-1: Map: 1 Cumulative CPU: 3.09 sec HDFS Read: 16156295 HDFS Write: 16150340
HDFS EC Read: 0 SUCCESS
[...]
```

使用 head 命令查看导出数据文件的前几行。

```
head /root/tmp/flights/*
2014,1,1,914,14,1238,13,0,AA,N338AA,1,JFK,LAX,359,2475,9,14
2014,1,1,1157,-3,1523,13,0,AA,N335AA,3,JFK,LAX,363,2475,11,57
2014,1,1,1902,2,2224,9,0,AA,N327AA,21,JFK,LAX,351,2475,19,2
2014,1,1,722,-8,1014,-26,0,AA,N3EHAA,29,LGA,PBI,157,1035,7,22
2014,1,1,1347,2,1706,1,0,AA,N319AA,117,JFK,LAX,350,2475,13,47
2014,1,1,1824,4,2145,0,0,AA,N3DEAA,119,EWR,LAX,339,2454,18,24
2014,1,1,2133,-2,37,-18,0,AA,N323AA,185,JFK,LAX,338,2475,21,33
2014,1,1,1542,-3,1906,-14,0,AA,N328AA,133,JFK,LAX,356,2475,15,42
2014,1,1,1509,-1,1828,-17,0,AA,N5FJAA,145,JFK,MIA,161,1089,15,9
2014,1,1,1848,-2,2206,-14,0,AA,N3HYAA,235,JFK,SEA,349,2422,18,48
```

2. 导出到 HDFS

使用 INSERT OVERWRITE DIRECTORY '<导出路径>'语句将 flights.flights14 表导出到 HDFS 上的数据文件/user/root/flights 中。

```
INSERT OVERWRITE DIRECTORY '/user/root/flights'
ROW FORMAT DELIMITED FIELDS TERMINATED BY ','
SELECT * FROM flights.flights14;
[...]
INFO   : Hadoop job information for Stage-1: number of mappers: 1; number of reducers: 0
INFO   : 2019-01-22 14:37:37,276 Stage-1 map = 0%,   reduce = 0%
INFO   : 2019-01-22 14:37:47,893 Stage-1 map = 100%,   reduce = 0%, Cumulative CPU 2.91 sec
[...]
INFO   : Starting task [Stage-0:MOVE] in serial mode
INFO   : Moving data to directory /user/root/flights from hdfs://master:8020/tmp_hive_2019-01-22_
14-37-24_013_3985870651674793883-1/-ext-10000
INFO   : MapReduce Jobs Launched:
INFO   : Stage-Stage-1: Map: 1 Cumulative CPU: 2.91 sec HDFS Read: 16156202 HDFS Write: 16150340
HDFS EC Read: 0 SUCCESS
[...]
```

使用 hdfs dfs -cat 命令查看导出数据文件的前几行。

```
hdfs dfs -cat flights/* | head
2014,1,1,914,14,1238,13,0,AA,N338AA,1,JFK,LAX,359,2475,9,14
2014,1,1,1157,-3,1523,13,0,AA,N335AA,3,JFK,LAX,363,2475,11,57
2014,1,1,1902,2,2224,9,0,AA,N327AA,21,JFK,LAX,351,2475,19,2
2014,1,1,722,-8,1014,-26,0,AA,N3EHAA,29,LGA,PBI,157,1035,7,22
2014,1,1,1347,2,1706,1,0,AA,N319AA,117,JFK,LAX,350,2475,13,47
2014,1,1,1824,4,2145,0,0,AA,N3DEAA,119,EWR,LAX,339,2454,18,24
2014,1,1,2133,-2,37,-18,0,AA,N323AA,185,JFK,LAX,338,2475,21,33
2014,1,1,1542,-3,1906,-14,0,AA,N328AA,133,JFK,LAX,356,2475,15,42
2014,1,1,1509,-1,1828,-17,0,AA,N5FJAA,145,JFK,MIA,161,1089,15,9
2014,1,1,1848,-2,2206,-14,0,AA,N3HYAA,235,JFK,SEA,349,2422,18,48
```

小　　结

本章主要介绍了 Hive 数据仓库的数据操作，包括数据导入、数据插入和数据导出。数据导入和导出的目标可以是本地文件系统或 HDFS 分布式文件系统。

对于 Hive 独有的分区表特性，插入需要使用特殊语法。一种为静态方式，即直接指定插入的分区；另一种为动态方式，即根据分区列的值动态决定插入的分区，而在此之前需要配置某些选项。

习　　题

1. Hive 从本地文件系统和从 HDFS 导入数据的语法有哪些区别？
2. 在 Hive 中执行从本地文件系统和从 HDFS 导入数据等操作。
3. Hive 中动态插入分区表需要配置哪些选项？
4. 在 Hive 中执行多次基本插入和插入并覆盖操作，观察数据表的记录条数区别。
5. 在 Hive 中执行动态插入数据表操作。
6. 在 Hive 中执行导出数据到本地文件系统和从 HDFS 等操作。

第 8 章

Hive 数据查询 «‹

本章所有例子同样基于航班的准点情况数据以及航空公司简写与描述的映射数据，之前章节已有介绍。数据集的路径为～/data/flights/flights14.csv 以及/root/data/flights/carriers.csv，其中第一行为列名。同时也已经存储在 Hive 的 flights.flights14 表以及 flights.carrier 表中。

8.1 基本查询

数据查询语句的基本语法和常用选项如下：

```
SELECT [DISTINCT] <列名表达式 1> [AS <列别名 1>], <列名表达式 2> [AS <列别名 2>], ...
FROM <表名>
[WHERE <筛选条件>]
[GROUP BY <列名表达式列表>]
[HAVING <筛选条件>]
[ORDER BY <列名表达式列表>]
[LIMIT <行数>]
```

使用 SELECT…FROM…LIMIT…语句从航班表 flights.flights14 中选取所有列，并返回前 5 行，其中：

- 星号（*）表示选取所有列；
- LIMIT 语句表示保留前几行。

```
SELECT *
FROM flights.flights14
LIMIT 5;
```

使用 WHERE 语句筛选航班表 flights.flights14 中出发地（列 origin）为'JFK'、月份（列 month）为 6 月的数据行，其中：

- 字符串可以使用单引号（'）或双引号（"）；
- 一个等号（=）表示比较是否相等；
- 运算符 and 表示多个条件“与”的关系。

```
SELECT *
FROM flights.flights14
WHERE origin = 'JFK' and month = 6
LIMIT 5;
```

```
+----------------+------------------+---------------+---------------------+----------------------+-----------------
------+---------------------+---------------------+--------------------+--------------------+-----------------+----
-------------------+----------------+--------------------+--------------------+----------------+----------------+--+
| flights14.year | flights14.month | flights14.day | flights14.dep_time | flights14.dep_delay | flights14.arr_
time | flights14.arr_delay | flights14.cancelled | flights14.carrier | flights14.tailnum+ | flights14.flight |
flights14.origin | flights14.dest | flights14.air_time | flights14.distance | flights14.hour | flights14.min |
+------+--+--+------+----+-------+----+--+----+----------+----+-----+-------+-----+------+----+---+--+
| 2014 | 6 | 1 |  851 |  -9 | 1205 |  -5 | 0 | AA | N787AA |  1 | JFK | LAX | 324 | 2475 |  8 | 51 |
| 2014 | 6 | 1 | 1220 | -10 | 1522 | -13 | 0 | AA | N795AA |  3 | JFK | LAX | 329 | 2475 | 12 | 20 |
| 2014 | 6 | 1 |  718 |  18 | 1014 |  -1 | 0 | AA | N784AA |  9 | JFK | LAX | 326 | 2475 |  7 | 18 |
| 2014 | 6 | 1 | 1024 |  -6 | 1314 | -16 | 0 | AA | N791AA | 19 | JFK | LAX | 320 | 2475 | 10 | 24 |
| 2014 | 6 | 1 | 1841 |  -4 | 2125 | -45 | 0 | AA | N790AA | 21 | JFK | LAX | 326 | 2475 | 18 | 41 |
+------+--+--+------+----+-------+----+--+----+----------+----+-----+-------+-----+------+----+---+--+
```

使用 ORDER BY 语句将返回结果按出发地（列 origin）顺序、目的地（列 dest）倒序排列。其中，DESC 关键字表示倒序排列，ASC 关键字表示顺序排列，默认为顺序排列，因此 ASC 可以省略。

```
SELECT *
FROM flights.flights14
ORDER BY origin, dest DESC
LIMIT 5;
+----------------+------------------+---------------+---------------------+----------------------+-----------------
------+---------------------+---------------------+--------------------+--------------------+-----------------+----
-------------------+----------------+--------------------+--------------------+----------------+----------------+--+
| flights14.year | flights14.month | flights14.day | flights14.dep_time | flights14.dep_delay | flights14.arr_
time | flights14.arr_delay | flights14.cancelled | flights14.carrier | flights14.tailnum+ | flights14.flight |
flights14.origin | flights14.dest | flights14.air_time | flights14.distance | flights14.hour | flights14.min |
+------+---+----+-----+---+----+-----+--+----+----------+------+-------+-------+-----+------+--+----+--+
| 2014 | 10 |  1 | 607 | -4 | 806 | -20 | 0 | EV | N34110 | 4419 | EWR | XNA | 160 | 1131 | 6 |  7 |
| 2014 | 10 | 31 | 634 | -2 | 834 | -25 | 0 | EV | N17138 | 4419 | EWR | XNA | 160 | 1131 | 6 | 34 |
| 2014 |  7 | 16 | 630 |  0 | 846 |   6 | 0 | EV | N11181 | 4419 | EWR | XNA | 160 | 1131 | 6 | 30 |
| 2014 |  5 | 13 | 635 |  0 | 836 | -25 | 0 | EV | N16183 | 4419 | EWR | XNA | 161 | 1131 | 6 | 35 |
| 2014 | 10 |  9 | 606 | -5 | 814 | -12 | 0 | EV | N24103 | 4419 | EWR | XNA | 166 | 1131 | 6 |  6 |
+------+---+----+-----+---+----+-----+--+----+----------+------+-------+-------+-----+------+--+----+--+
```

使用 SELECT 语句后跟列名，选取航班到达延误分钟数（列 arr_delay）和出发延误分钟数（列 dep_delay），并使用 AS 关键字将这两列重命名为 delay_arr 和 delay_dep。

```
SELECT arr_delay AS delay_arr, dep_delay AS delay_dep
FROM flights.flights14
LIMIT 5;
+------------+-------------+--+
| delay_arr  | delay_dep   |
+------------+-------------+--+
| 13         | 14          |
| 13         | -3          |
|  9         |  2          |
| -26        | -8          |
|  1         |  2          |
+------------+-------------+--+
```

使用 SELECT 语句后跟列名的表达式，重新计算返回 2 列：平均速度（列 speed）和总延误分钟数（列 delay）。

```
SELECT distance / (air_time / 60) AS speed,
arr_delay + dep_delay AS delay
FROM flights.flights14
LIMIT 5;
+--------------------------+----------+
|          speed           | delay    |
+--------------------------+----------+
| 413.6490250696379        | 27       |
| 409.0909090909091        | 10       |
| 423.0769230769231        | 11       |
| 395.54140127388536       | -34      |
| 424.28571428571433       | 3        |
+--------------------------+----------+
```

在 SELECT 语句中，使用 DISTINCT 关键字表示仅选取独立值。

```
SELECT DISTINCT origin
FROM flights.flights14;
+-----------+--+
|  origin   |
+-----------+--+
| EWR       |
| JFK       |
| LGA       |
+-----------+--+
```

8.2 数据聚合

所谓数据聚合，就是将原数据中的多行按一定的计算方法合并成一行返回，其中有两个要素：

- 计算方法：在 Hive 中通过聚合函数指定；
- 分组方式：即哪些原数据中的行合并成返回结果的一行，在 Hive 中通过 SELECT 语句中未包含在聚合函数中的列以及 GROUP BY 语句中的列指定。

调用聚合函数 count()计算 flights.flights14 表的行数。

```
SELECT count(1)
FROM flights.flights14;
+-----------+--+
|    _c0    |
+-----------+--+
| 253316    |
+-----------+--+
```

使用 GROUP BY 语句指定按出发地（列 origin）分组，计算各出发地的行数。

```
SELECT origin, count(1)
FROM flights.flights14
```

```
GROUP BY origin;
+---------+----------+--+
| origin  |   _c1    |
+---------+----------+--+
| EWR     | 87400    |
| JFK     | 81483    |
| LGA     | 84433    |
+---------+----------+--+
```

使用 GROUP BY 语句指定多个列，计算各不同出发地（列 origin）和目的地（列 dest）的组合中，航空公司代码（列 carrier）为'AA'的行数。

```
SELECT origin, dest, count(1)
FROM flights.flights14
WHERE carrier = 'AA'
GROUP BY origin, dest;
+------------+----------+-------+--+
| origin     | dest     |  _c2  |
+------------+----------+-------+--+
| EWR        | DFW      | 1618  |
| EWR        | LAX      | 62    |
| EWR        | MIA      | 848   |
| EWR        | PHX      | 121   |
| JFK        | AUS      | 297   |
| JFK        | BOS      | 1173  |
| JFK        | DCA      | 172   |
| JFK        | DFW      | 474   |
| JFK        | EGE      | 85    |
| JFK        | IAH      | 7     |
| JFK        | LAS      | 595   |
| JFK        | LAX      | 3387  |
| JFK        | MCO      | 597   |
| JFK        | MIA      | 1876  |
| JFK        | ORD      | 432   |
| JFK        | SAN      | 299   |
| JFK        | SEA      | 298   |
| JFK        | SFO      | 1312  |
| JFK        | SJU      | 690   |
| JFK        | STT      | 229   |
| LGA        | DFW      | 3785  |
| LGA        | MIA      | 3334  |
| LGA        | ORD      | 4366  |
| LGA        | PBI      | 245   |
+------------+----------+-------+--+
```

使用 HAVING 语句对聚合结果做进一步筛选，返回所有出发地（列 origin）和目的地（列 dest）组合中，航空公司代码（列 carrier）为'AA'的行数大于 3000 的记录。

```
SELECT origin, dest, count(1)
FROM flights.flights14
WHERE carrier = 'AA'
```

```
GROUP BY origin, dest
HAVING count(1) > 3000;
+------------+-----------+--------+
|  origin   |  dest   |  c2   |
+------------+-----------+--------+
| JFK       | LAX     | 3387  |
| LGA       | DFW     | 3785  |
| LGA       | MIA     | 3334  |
| LGA       | ORD     | 4366  |
+------------+-----------+--------+
```

Hive 有如下聚合函数实现常用统计功能，如表 8-1 所示。

表 8-1 常用聚合函数

函　　数	功　　能
count()	非空元素个数
sum()	总和
max()和 min()	最大和最小的元素
avg()	均值
var_pop()和 var_samp()	总体和样本方差
stddev_pop()和 stddev_samp()	总体和样本标准差
covar_pop()和 covar_samp()	总体和样本协方差
corr()	皮尔森相关性系数
percentile()	百分位数

```
SELECT origin,
count(1) as count,
sum(dep_delay) AS sum_dep_delay,
max(dep_delay) AS max_dep_delay,
min(dep_delay) AS min_dep_delay,
avg(dep_delay) AS avg_dep_delay,
var_pop(dep_delay) AS var_dep_delay,
stddev_pop(dep_delay) AS stddev_dep_delay,
covar_pop(dep_delay, arr_delay) AS covar_dep_delay,
corr(dep_delay, arr_delay) AS corr_dep_delay,
percentile(dep_delay, array(0.01,0.1,0.25,0.5,0.75,0.9,0.99)) AS percentile_dep_delay
FROM flights.flights14
GROUP BY origin;
+---------+--------+-------------------+------------------+------------------+------------------------+---------
-----------------+----------------------+---------------------+-------------------+----------------------+--+
| origin  | count  | sum_dep_delay  | max_dep_delay  | min_dep_delay  |     avg_dep_delay     |
var_dep_delay  |  stddev_dep_delay  |  covar_dep_delay  |  corr_dep_delay  | percentile_dep_delay |
+-------+-------+-----------+-----+-----+-----------------------+-----------------------+----------------------+-
-------------------------+---------------------------+------------------------------------------+--+
| EWR | 87400 | 1329571 | 1498 | -27 | 15.212482837528604 | 1862.4487526455873 | 43.15609751408933 |
1899.4167380055253   | 0.934393844230353     | [-11.0,-7.0,-4.0,0.0,15.0,54.0,202.0]   |
| JFK  | 81483 | 932668   | 1241 | -34 | 11.446166685075415 | 1714.5489808937655 | 41.40711268482464 |
1754.8203622011424   | 0.9166548197372786    | [-10.0,-7.0,-5.0,-2.0,9.0,40.0,197.0]   |
```

```
| LGA | 84433 | 895412   | 973  |-112 | 10.605000414529863 | 1609.371683687964  | 40.11697500669715 |
1611.7267330182738   | 0.9183727851799162   | [-13.0,-9.0,-6.0,-3.0,8.0,44.0,186.0]   |
+-------+-------+-----------+-----+-----+------------------------+------------------------+------------------------+-
--------------------------+---------------------------+---------------------------------------------+--+
```

8.3 数据连接

数据连接是指两个数据集按指定的连接条件按列合并成一个数据集，如图 8-1 所示。主要分为以下几种：

- 内连接（inner join）：包含左侧数据集和右侧数据集中都存在且满足连接条件的行；
- 左连接（left join）：包含左侧数据集中所有行，右侧数据集中不满足连接条件的列设为空；
- 右连接（right join）：包含右侧数据集中所有行，左侧数据集中不满足连接条件的列设为空；
- 全连接（full join）：包含左侧数据集和右侧数据集中所有行，不满足连接条件的列设为空。

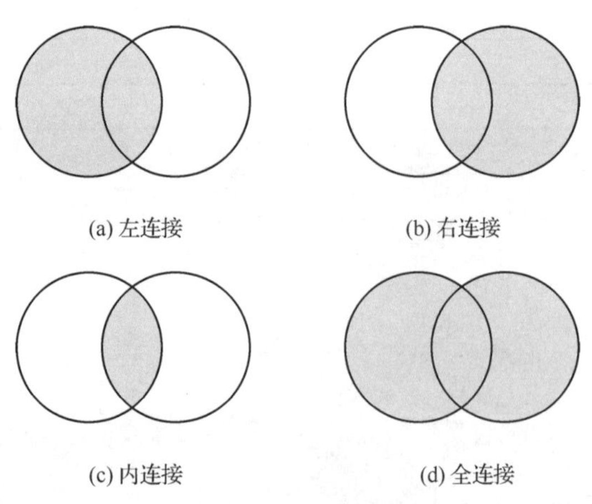

图 8-1　数据连接的种类

Hive 中的 JOIN 语句用于做数据连接，具体为：

- <左侧表名> [INNER] JOIN <右侧表名> ON <连接条件>表示内连接，其中 INNER 关键字可以省略；
- <左侧表名> LEFT | RIGHT | FULL [OUTER] JOIN <右侧表名> ON <连接条件> 表示左连接、右连接和全连接，其中 OUTER 关键字可以省略。

1. 内连接

使用 INNER JOIN 命令做内连接，并指定连接条件为航空公司代码（列 carrier）相等。

需要注意的是，表名后面可以加上表的别名，在指定表的列时，可以使用<表别名>.<列名>的形式。

```
DROP TABLE IF EXISTS flights_carrier;
CREATE TABLE IF NOT EXISTS flights_carrier AS
SELECT a.*, b.description
FROM flights.flights14 a
INNER JOIN flights.carrier b
ON a.carrier = b.carrier;
```

计算内连接后的记录条数。

```
SELECT count(1) FROM flights_carrier;
+-----------+--+
|    _c0    |
+-----------+--+
| 235315    |
+-----------+--+
```

可以看出，连接后的记录条数少于连接前航班数据集的记录条数，意味着航班数据集中有一些记录并没有出现在结果中。

2. 左连接

使用 LEFT JOIN 命令做左连接，并指定连接条件为航空公司代码（列 carrier）相等。

```
DROP TABLE IF EXISTS flights_carrier;
CREATE TABLE IF NOT EXISTS flights_carrier AS
SELECT a.*, b.description
FROM flights.flights14 a
LEFT JOIN flights.carrier b
ON a.carrier = b.carrier;
```

计算左连接后的记录条数。

```
SELECT count(1) FROM flights_carrier;
+-----------+--+
|    _c0    |
+-----------+--+
| 253316    |
+-----------+--+
```

可以看出，连接后的记录条完全等于连接前航班数据集的记录条数，意味着航班数据集中有全部记录都出现在结果中。

选取连接结果中航空公司描述（列 description）为空的记录。

```
SELECT *
FROM flights_carrier
WHERE description IS NULL
LIMIT 5;
+----------------------+-------------------------+----------------------+---------------------------+----------
----------------------+----------------------------+-------------------------------+-------------------------------+
---------------------------+----------------------------+------------------------------+------------------------------+
```

```
--------------------------+-------------------------------+---------------------------------+---------------------------+
-------------------------+---------------------------------+--+
| flights_carrier.year | flights_carrier.month | flights_carrier.day | flights_carrier.dep_time | flights_carrier.dep_delay | flights_carrier.arr_time | flights_carrier.arr_delay | flights_carrier.cancelled | flights_carrier.carrier | flights_carrier.tailnum | flights_carrier.flight | flights_carrier.origin | flights_carrier.dest | flights_carrier.air_time | flights_carrier.distance | flights_carrier.hour | flights_carrier.min | flights_carrier.description |
+------+--+--+------+----+-------+----+--+---+-----------+------+------+-------+-----+----+----+---+--------+--+
| 2014 | 1 | 1 | 1146 |  -5 | 1409 | -10 | 0 | FL | N952AT | 63   | LGA | ATL | 124 | 762 | 11 | 46 | NULL |
| 2014 | 1 | 1 | 945  |  -5 | 1117 |  -6 | 0 | FL | N922AT | 160  | LGA | CAK | 72  | 397 | 9  | 45 | NULL |
| 2014 | 1 | 1 | 936  | 16  | 1201 |  13 | 0 | FL | N982AT | 281  | LGA | ATL | 123 | 762 | 9  | 36 | NULL |
| 2014 | 1 | 1 | 1734 |   5 | 1959 |   4 | 0 | FL | N980AT | 400  | LGA | ATL | 129 | 762 | 17 | 34 | NULL |
| 2014 | 1 | 1 | 1421 |  -4 | 1647 |  -6 | 0 | FL | N996AT | 1070 | LGA | ATL | 128 | 762 | 14 | 21 | NULL |
+------+--+--+------+----+-------+----+--+---+-----------+------+------+-------+-----+----+----+---+--------+--+
```

右连接与左连接类似，在此不再赘述。

3. 全连接

使用 FULL JOIN 命令做全连接，并指定连接条件为航空公司代码（列 carrier）相等。

```
DROP TABLE IF EXISTS flights_carrier;
CREATE TABLE IF NOT EXISTS flights_carrier AS
SELECT a.*, b.description
FROM flights.flights14 a
FULL JOIN flights.carrier b
ON a.carrier = b.carrier;
```

计算全连接后的记录条数。

```
SELECT count(1) FROM flights_carrier;
+-----------+--+
|    _c0    |
+-----------+--+
| 253728    |
+-----------+--+
```

可以看出，连接后的记录条数大于连接前航班数据集的记录条数，意味着航空公司代码数据集中的列 carrier 的有些数值没有出现在航班数据集中。

小　　结

本章主要介绍了 Hive 数据仓库的数据查询操作。基本查询中包含 SELECT、FROM、WHERE、GROUP BY、HAVING、ORDER BY、LIMIT 等子句的具体用法。在数据聚合操作中，介绍了基本用法和常用的聚合函数。最后，介绍了数据连接的几种方式和基本用法，包括内连接、左（右）连接和全连接。数据连接是指两个数据集按指定的连接条件按列合并成一个数据集。

习　题

1. Hive数据库的操作语法与其他学过的数据库语法有什么不同？尝试总结。

2. 内连接、左连接和全连接有什么不同？

3. 基于航班准点数据集，哪个月份的平均到达延误分钟数最大？

4. 基于航班准点数据集，哪个航空公司（名称）的飞机数量（通过不同的尾翼编号确定）最多？

Spark 数据分析

第9章

Spark 简介 «‹

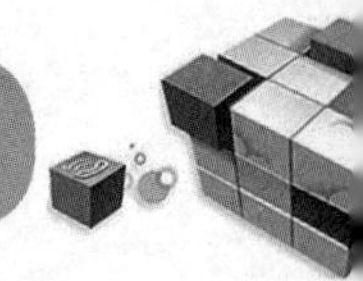

9.1　Spark 概述

Spark 是一个通用、高效的集群计算框架，用于处理海量数据，被业界广泛认为是 MapReduce 计算框架的继承者。Spark 不同于 Hive，并不依赖于 MapReduce 作为执行引擎，而是基于自身的执行库。诚然，Spark 与 MapReduce 在调用接口和运行方式上有许多相似之处。同时，Spark 与 Hadoop 集成紧密，Spark 应用可以在 YARN 上运行和调度，Spark 支持 HDFS 文件存储。

Spark 能够将属于同一个应用中不同作业之间的中间结果数据集保存在内存中，而 MapReduce 需要将每次作业的输出写入磁盘。两类应用的效率能够从这一特性中得到显著提升。第一类应用是迭代式的算法，即在同一个数据集上重复执行一系列操作直到满足停止条件。多数机器学习算法为迭代式算法，每次迭代会调整模型参数以优化目标函数，当目标函数满足一定条件时则停止迭代。官方公布的 Spark 和 MapReduce 运行逻辑回归所需时间如图 9-1 所示。第二类应用是交互式数据分析，即用户多次对同一个数据集做不同的查询。

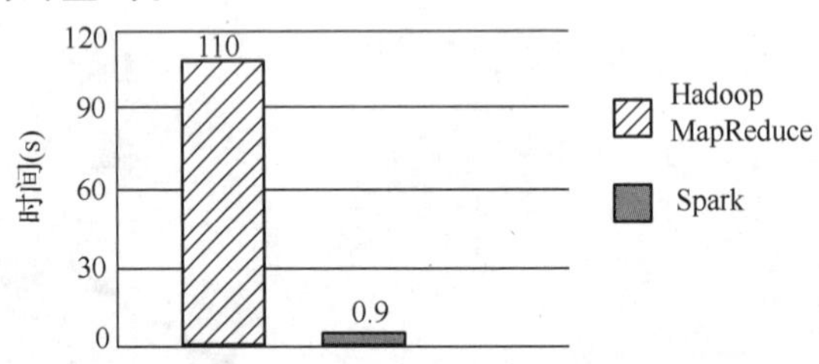

图 9-1　Spark 和 MapReduce 运行逻辑回归时间图

Spark 目前支持 4 种语言的 API，包括 Scala、Java、Python 和 R。本书中的例子主要用 Python 编写。Spark 还提供了一系列高层的数据分析函数库，例如：

- MLlib 用于机器学习；
- GraphX 用于图的处理；
- Spark Streaming 用于流处理；
- Spark SQL 用于数据框和 SQL 操作。

本书重点介绍其中的 MLlib 和 Spark SQL。

9.2 Spark 原理

Spark 有应用（application）、作业（job）、阶段（stage）和任务（task）的概念。每个 Spark 应用都会对应一个 SparkSession 实例（后续章节介绍），一个应用可以是一个或多个作业，串行或并行执行，同一个应用的不同作业之间可以共享缓存在内存中的数据集。Spark 作业比 MapReduce 作业更通用，包含了多个阶段的有向无环图（directed acyclic graph，DAG），每个阶段类似于 Map 阶段或 Reduce 阶段。一个阶段包含了若干并行化的任务，分配到不同节点执行。

一个 Spark 应用在执行中分为一个驱动器（driver）和若干个分布在集群中不同节点的执行器（executor）。

驱动器中有 SparkSession 实例，用于维护应用状态，存在于整个应用程序的生命周期，并连接 YARN 资源管理器，进一步向 YARN 申请集群资源。根据提交应用时的选项，驱动器可以是提交应用的客户端进程，也可以存在于客户端进程不同的集群节点中，后续章节会具体介绍。在交互式模式中，Spark Shell 进程本身就是驱动器。

执行器用于执行应用中具体的数据处理任务，同时用于缓存数据。驱动器将应用代码发送给执行器，并在其中执行任务，一个执行器可以同时执行多个任务。事实上，Spark 应用不仅可以在 YARN 上运行，还可以在其他集群管理器（如 Mesos 等）上运行，本书仅以 YARN 为例。应用执行的具体流程如图 9-2 所示。执行器的生命周期取决于动态分配机制是否启动。如果启动（默认配置），则当应用有等待任务时，会请求更多的执行器，当应用闲置时，执行器会被释放。

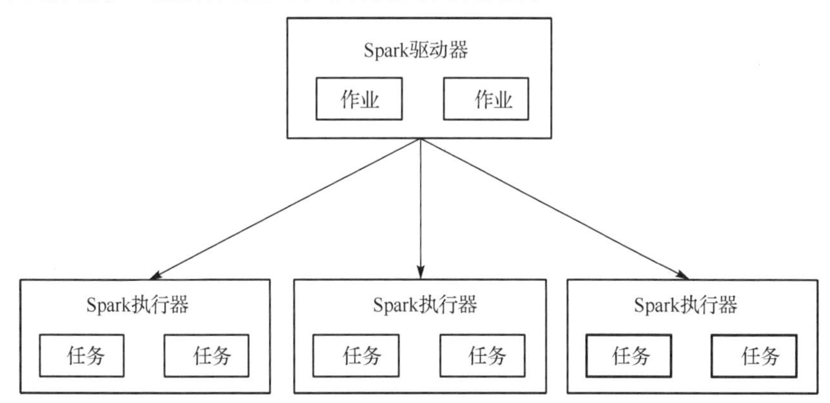

图 9-2　Spark 应用执行流程图

简言之，Spark 应用的运行包括以下几个步骤：

① 驱动器启动，并开始执行主函数；

② 驱动器向集群管理器请求资源，用于运行执行器；

③ 集群管理器启动执行器；

④ 驱动器运行应用，根据应用中的操作，将任务发送给执行器执行；

⑤ 任务在执行器中运行，并保存结果；

⑥ 如果启动动态分配机制，则当执行器闲置一段时间后，将被释放；

⑦ 应用的主函数退出或调用 SparkSession.stop()函数时，释放所有执行器。

9.3 Spark 的安装

1．安装 Spark（所有节点）

使用 yum install 命令在线安装 Spark 相关组件，包括：

- Spark 核心服务：spark-core.noarch；
- Spark 历史服务：spark-history-server.noarch；
- Spark Python 客户端：spark-python.noarch。

```
sudo yum install spark-core.noarch
sudo yum install spark-history-server.noarch
sudo yum install spark-python.noarch
```

2．配置 Spark（主节点）

打开 Spark 配置文件/etc/spark/conf/spark-defaults.conf，加入的配置项包括：

- spark.driver.memory：Spark 驱动器所占用的内存，仅在集群模式有效；
- spark.yarn.am.memory：Spark 应用主程序所占用的内存，仅在客户端部署模式有效；
- spark.executor.memory：Spark 执行器所占用的内存；
- spark.driver.memoryOverhead：Spark 驱动器所占用的除了堆（heap）以外的额外内存；
- spark.executor.memoryOverhead：Spark 执行器所占用的除了堆（heap）以外的额外内存；

在该文件末尾部分添加以下几行。

```
vim /etc/spark/conf/spark-defaults.conf
spark.driver.memory 512m
spark.yarn.am.memory 512m
spark.executor.memory 512m
spark.driver.memoryOverhead 128m
spark.executor.memoryOverhead 128m
```

打开 Spark 从节点列表配置文件/etc/hadoop/conf.my_cluster/workers，加入从节点的主机名。

```
vim /etc/spark/conf/slaves
```

该文件看起来如下所示。

```
slave1
slave2
```

将 master 上的 slaves 文件复制到主机 slave1 和主机 slave2 上

使用 scp 命令从主节点复制配置文件路径到从节点的相同路径。

```
scp -r /etc/spark/conf/   slave1:/etc/spark/
scp -r /etc/spark/conf/   slave1:/etc/spark/
```

3. 建立/user/spark 文件夹（主节点）

Hadoop 中的许多组件在运行时会在 HDFS 上生成许多临时文件存储在/tmp 文件夹中。

Spark 应用历史服务在运行时会将一些数据存储在 HDFS 上的/user/spark/applicationHistory 文件夹中。

以 hdfs 用户使用 hdfs dfs –mkdir 命令创建文件夹/user/spark/applicationHistory 以及 hdfs dfs –chown 命令将文件夹所有权移交给 spark 用户。

```
sudo -u hdfs hdfs dfs -mkdir -p /user/spark/applicationHistory
sudo -u hdfs hdfs dfs -chown -R spark /user/spark
```

9.4 Spark 运行方式

Spark 有两种运行方式：交互式的 Spark Shell 和独立应用（self–contained application）。

1. 交互分析

在交互式的 Spark Shell 中，用户输入命令并实时返回结果，用于学习 Spark 的 API 或探索性数据分析，支持 Python 和 Scala，本书中的所有例子都可以使用这种方式运行，且以 Python 撰写。

打开控制台，使用 pyspark 命令启动 Python 版本的 Spark Shell。

```
pyspark
Python 2.7.5 (default, Oct 30 2018, 23:45:53)
[GCC 4.8.5 20150623 (Red Hat 4.8.5-36)] on linux2
Type "help", "copyright", "credits" or "license" for more information.
[...]
Welcome to
      ____              __
     / __/__  ___ _____/ /__
    _\ \/ _ \/ _ `/ __/  '_/
   /__ / .__/\_,_/_/ /_/\_\   version 2.4.0-cdh6.1.0
      /_/

Using Python version 2.7.5 (default, Oct 30 2018 23:45:53)
SparkSession available as 'spark'.
```

这里的 spark 对象是 SparkSession 类的一个实例，用于维护应用状态，存在于整个应用程序的生命周期，并连接 YARN 资源管理器，进一步向 YARN 申请集群资源。

这里直接输入一段计算 π 的代码，具体细节会在之后章节做解释，基本步骤如下：

① 定义变量 partitions 为数据分区数；

② 定义变量 n 为生成随机点的个数；

③ 定义函数 f()，其中随机生成变量 x 和 y 为[–1.0,1.0)的随机数，判断以变量 x 为横轴、变量 y 为纵轴的点距离原点的距离是否小于 1，是则返回 1，否则返回 0；

④ 统计生成的随机点中，与原点距离小于 1 的点的个数，赋给变量 count；

⑤ π 约等于 4 倍的变量 count 与 n 之比。

```
from random import random
from operator import add

partitions = 2
n = 100000 * partitions

def f(_):
    x = random() * 2 - 1
    y = random() * 2 - 1
    return 1 if x ** 2 + y ** 2 < 1 else 0

count = spark.sparkContext.parallelize(range(1, n + 1), partitions). map(f).reduce(add)
print("Pi is roughly %f" % (4.0 * count / n))
Pi is roughly 3.141100
```

调用 exit()函数退出 Spark Shell。

```
exit()
```

2. 独立应用

独立应用程序是一种非交互使用方式，即将所有数据处理过程写在一个代码文件中，整体提交给 Spark，支持 Python、Scala、Java 和 R。

这里直接调用现成的计算 π 的代码文件，流程类似。

```
import sys
from random import random
from operator import add

from pyspark.sql import SparkSession

if __name__ == "__main__":
    """
        Usage: pi [partitions]
    """
    spark = SparkSession\
        .builder\
        .appName("PythonPi")\
        .getOrCreate()

    partitions = int(sys.argv[1]) if len(sys.argv) > 1 else 2
    n = 100000 * partitions

    def f(_):
        x = random() * 2 - 1
        y = random() * 2 - 1
        return 1 if x ** 2 + y ** 2 <= 1 else 0

    count = spark.sparkContext.parallelize(range(1, n + 1), partitions).map(f).reduce(add)
    print("Pi is roughly %f" % (4.0 * count / n))

    spark.stop()
```

使用 spark-submit 命令提交代码文件启动独立应用。

```
# tar xf /usr/lib/spark/python.tar.gz
spark-submit /usr/lib/spark/pi.py
...
Pi is roughly 3.141840
```

9.5 Spark 运行位置

运行交互式的 Spark Shell 或独立应用时，都可以在启动 Spark 时指定一系列运行参数。如运行 Spark Shell 时，可通过以下形式指定运行参数。

```
pyspark --<选项 1> <值 1> --<选项 2> <值 2> ...
```

这里首先介绍选项--master，表示应用运行的位置，常用的值包括：

- local 表示在本地运行单个工作线程的 Spark，即没有任何并行；
- local[K]表示在本地运行 K 个工作线程的 Spark，理想状态下可以将 K 设为本地 CPU 核数；
- local[*]表示在本地运行与本地 CPU 核数相同个工作线程的 Spark；
- yarn 表示在 YARN 上运行 Spark，为默认值。

其他选项会在后续章节介绍。

1. 在本地运行 Spark

打开控制台，使用 pyspark 命令时指定启动 2 个工作线程的 Spark。

```
pyspark --master local[2]
```

在其中再次运行计算 π 的代码。此时打开 YARN 资源管理器网页界面查看正在运行的应用（http://master:8088/cluster/apps/RUNNING），将看不到任何正在运行的 Spark 应用。因为该应用仅在本地运行，并没有提交到 YARN。

调用 exit()函数退出 Spark Shell。

2. 在 YARN 上运行 Spark

在 YARN 中，每个应用会有一个应用主程序（ApplicationMaster）作为应用的第一个容器（container）。应用主程序负责从资源管理器（ResourceManager）请求资源。当资源分配完毕后，应用主程序与节点管理器（NodeManager）通信以启动容器用于执行任务。启动 Spark 时，指定选项 master 为 yarn 表示在 YARN 上运行 Spark。

在 YARN 上运行 Spark 又分为服务器模式和客户端模式，取决于 Spark 驱动器处在应用主程序或客户端程序中。

在服务器模式中，Spark 驱动器运行在应用主程序中，因此应用主程序所在的容器既负责驱动应用，又负责向 YARN 请求运行执行器的资源。提交应用的客户端程序不需要在应用整个生命周期中都持续运行。具体流程如图 9-3 所示。通常情况下，服务器模式不适用于交互式的应用，因为这些应用需要驱动器运行在客户端程序中接受用户指令。

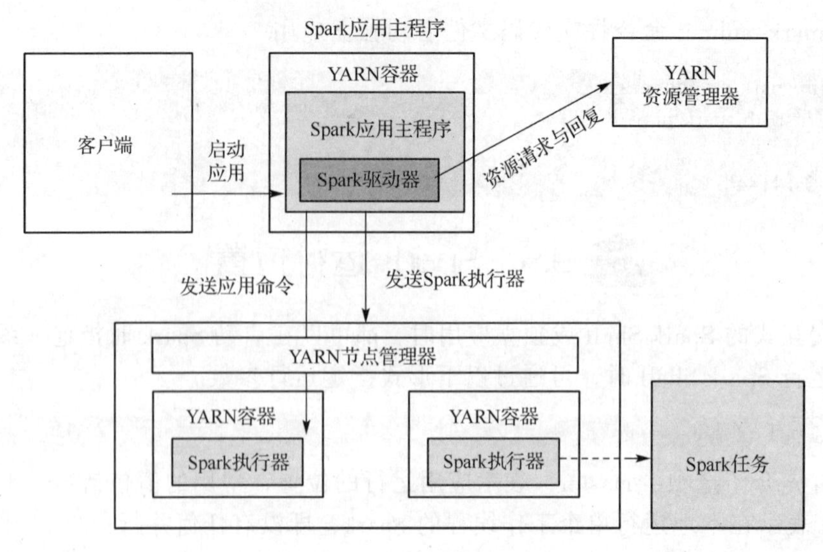

图 9-3　Spark 驱动器运行图

在客户端模式中，Spark 驱动器运行在客户端程序中，即提交应用的主机中。应用主程序仅负责向 YARN 请求运行执行器的资源。具体流程如图 9-4 所示。

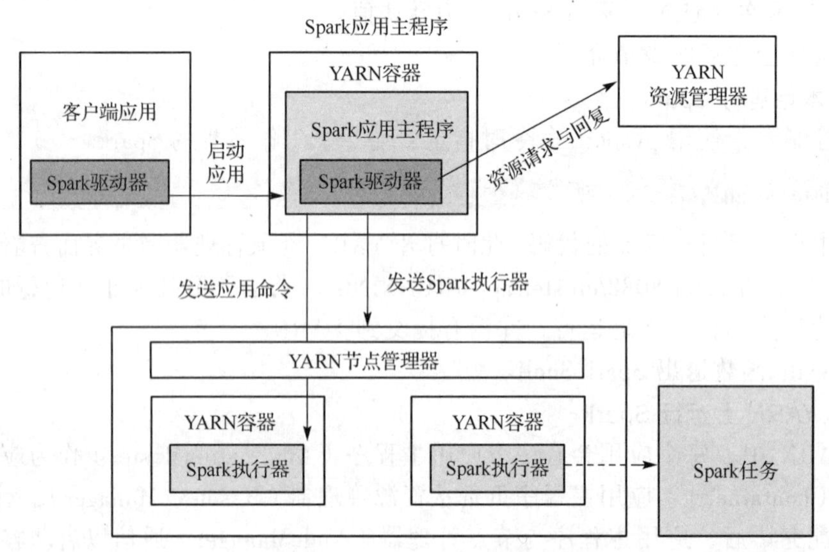

图 9-4　客户端 Spark 驱动器运行图

Spark 运行选项--deploy-mode 表示部署模式，其中：

- cluster 表示集群模式；
- client 表示客户端模式，为默认值。

打开控制台，使用 pyspark 命令时指定运行位置为 YARN，部署模式为客户端模式。

```
pyspark --master yarn --deploy-mode client
```

此时打开 YARN 资源管理器网页界面查看正在运行的应用，如图 9-5 所示。看到一个名为 PySparkShell 的应用，共占用 2 个容器，其中一个容器运行应用主程序，另一个容器运行执行器。

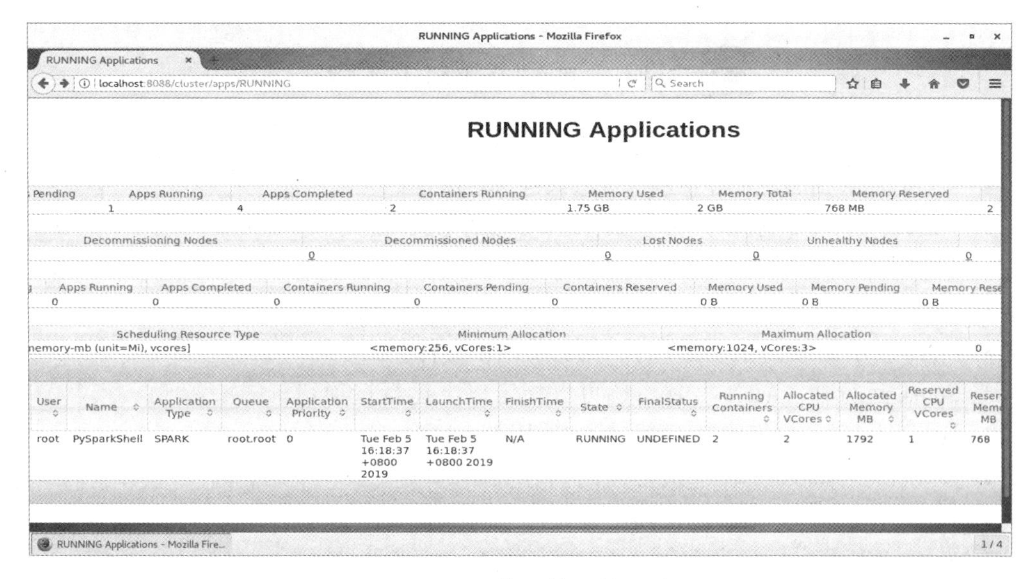

图 9-5 YARN 资源管理器网页界面

在其中再次运行计算 π 的代码。

调用 exit()函数退出 Spark Shell。

使用 spark-submit 命令启动计算 π 的独立程序，并指定运行位置为 YARN，部署模式为服务器模式。

```
spark-submit --master yarn --deploy-mode cluster /usr/lib/spark/pi.py
```

此时打开 YARN 资源管理器网页界面查看正在运行的应用，如图 9-6 所示。看到一个名为 pi.py 的应用，共占用 2 个容器，其中一个容器运行应用主程序，另一个容器运行执行器。

Applications Places Firefox Web Browser Tue 16:29
RUNNING Applications - Mozilla Firefox
RUNNING Applications
localhost:8088/cluster/apps/RUNNING
Search
RUNNING Applications
Pending | Apps Running | Apps Completed | Containers Running | Memory Used | Memory Total | Memory Reserved | VCo
1 | 5 | 2 | 1.50 GB | 2 GB | 768 MB | 2
Decommissioning Nodes | Decommissioned Nodes | Lost Nodes | Unhealthy Nodes | Re
0 | 0 | 0 | 0
Apps Running | Apps Completed | Containers Running | Containers Pending | Containers Reserved | Memory Used | Memory Pending | Memory Reserved
0 | 0 | 0 | 0 | 0 | 0 B | 0 B | 0 B
Scheduling Resource Type | Minimum Allocation | Maximum Allocation
emory-mb (unit=Mi), vcores] | <memory:256, vCores:1> | <memory:1024, vCores:3> | 0
User | Name | Application Type | Queue | Application Priority | StartTime | LaunchTime | FinishTime | State | FinalStatus | Running Containers | Allocated CPU VCores | Allocated Memory MB | Reserved CPU VCores | Reserved Memory MB
root | pi.py | SPARK | root.root | 0 | Tue Feb 5 16:28:10 +0800 2019 | Tue Feb 5 16:28:10 +0800 2019 | N/A | RUNNING | UNDEFINED | 2 | 2 | 1536 | 1 | 768
RUNNING Applications - Mozilla Fire...
1/4

图 9-6 YARN 资源管理器网页界面

9.6 Spark 运行参数

除了运行位置和部署模式，其他常用的 Spark 运行选项包括：

- --conf 表示 Spark 配置选项，形如<键>=<值>；
- --driver-cores 表示在集群模式中驱动器的 CPU 核数；
- --driver-memory 表示驱动器的内存大小；
- --files 表示需要放在执行器工作文件夹中的文件列表，以逗号分隔；
- --jars 表示加载到驱动器和执行器的 JAR 文件。

当 Spark 运行在 YARN 上时，还可以指定以下运行选项：

- --executor-cores 表示执行器的 CPU 核数；
- --executor-memory 表示执行器的内存大小；
- --num-executors 表示执行器的数量；
- --queue 表示应用提交到的 YARN 队列名。

以下例子使用 pyspark 命令时指定执行器内存大小为 128 MB,执行器数量为 2 个。

```
pyspark --executor-memory 128m --num-executors 2
```

调用 exit()函数退出 Spark Shell。

以下例子使用 pyspark 命令时指定加载 2 个 JAR 文件，用于读取 CSV 文件。

```
pyspark --jars spark-csv_2.10-1.5.0.jar,commons-csv-1.5.jar
```

调用 exit()函数退出 Spark Shell。

小　　结

本章首先介绍了 Spark 的产生背景和与 MapReduce 相比较的优势。Spark 能够将属于同一个应用中不同作业之间的中间结果数据集保存在内存中，而 MapReduce 需要将每次作业的输出写入磁盘。然后介绍了 Spark 的工作原理。一个 Spark 应用在执行中分为一个驱动器和若干个分布在集群中不同节点的执行器。

本章还着重介绍了 Spark 的安装流程，包括基础软件包的安装（Spark 核心服务、Spark 历史服务和 Spark Python 客户端）、Spark 的配置和验证安装结果。Spark 可以以交互式或独立应用的方式运行。

习　　题

1. Spark 与 MapReduce 有什么区别？
2. Spark 支持哪几种语言的 API?
3. Spark 的高层数据分析函数库有哪些？
4. Spark 的运行方式有哪些？各适用于什么场景？
5. 熟悉 Spark 的安装流程，搭建 Spark 环境。
6. 验证 Spark 安装是否成功，并运行测试程序。
7. 熟悉 Spark 的常用运行选项，并逐一尝试。

第10章 Spark 大数据处理

Spark SQL 是 Spark 中用于结构化数据处理的模块。不同于基本的基于 Spark RDD 的 API，Spark SQL 的 API 提供更多关于数据结构和执行计算的信息。Spark SQL 的执行引擎会利用这些额外信息作系统优化。Spark SQL 的接口又进一步分为 SQL 查询、数据框（DataFrame）API 和数据集（Dataset）API。无论使用这 3 种接口的哪一种，底层的执行引擎都是一致的，这意味着用户可以在这 3 种 API 之间无缝切换。本章主要介绍前两者。

Spark SQL 可以执行符合标准 SQL 语法或 HiveQL 的 SQL 查询，也可以直接从 Hive 中读取数据。数据框是一种分布式存储的数据集，由多个命名的列组成。数据框在理念上类似于关系型数据库中的数据表以及 R 语言和 Python 中的数据框，但在性能上更加优化。数据框可以从多个源头获取，如结构化的数据文件、Hive 数据表、外部数据库或已有的 RDD。

```
pyspark
```

载入本章需要用到的程序包。

```
from pyspark.sql import functions as F
from pyspark.sql.functions import avg, count, countDistinct, max, min, mean, variance, stddev, sum, skewness, kurtosis
```

本章所有例子同样基于航班的准点情况数据以及航空公司简写与描述的映射数据，之前章节已有介绍。数据集的路径为~/data/flights/flights14.csv 以及/root/data/flights/carriers.csv，其中第一行为列名。

10.1 数据框的创建

1. 通过键入创建

调用 spark 的 createDataFrame()函数创建数据框，其中：

- 第 1 个参数 data 表示输入数据，可以是 RDD、列表或 pandas.DataFrame 对象；
- 第 2 个参数 schema 表示表结构（列名和数据类型），当为列名的列表时，则自动推测数据类型，当为 None 时，则自动推测列名和数据类型。

以下例子从元组的列表创建数据框,其中第2个参数表示数据集的变量名即列名。

```
l = [('Nanjing','Jiangsu',8.335),('Hangzhou','Zhejiang',9.468), ('Fuzhou','Fujian',7.64)]
df = spark.createDataFrame(l,['city','province','population'])
```

调用数据框的 show()函数查看数据集前几行，默认为前 20 行。

```
df.show()
+------------+----------+--------------+
|    city    | province | population   |
+------------+----------+--------------+
|  Nanjing   | Jiangsu  |        8.335 |
| Hangzhou   |Zhejiang  |        9.468 |
|  Fuzhou    |   Fujian |         7.64 |
+------------+----------+--------------+
```

以下例子从字典的列表创建数据框，其中字典的键表示数据集的变量名即列名，字典的值表示每行每列的数据值。

```
d = [{'city': 'Nanjing', 'province': 'Jiangsu', 'population': 8.335},
     {'city': 'Hangzhou', 'province': 'Zhejiang', 'population': 9.468},
     {'city': 'Fuzhou', 'province': 'Fujian', 'population': 7.64}]
df = spark.createDataFrame(d)
df.show()
+------------+-------------+-----------+
|    city    | population  | province  |
+------------+-------------+-----------+
|  Nanjing   |       8.335 | Jiangsu   |
| Hangzhou   |       9.468 |Zhejiang   |
|  Fuzhou    |        7.64 |   Fujian  |
+------------+-------------+-----------+
```

以下例子从列表的 RDD 创建数据框，其中 SparkContext 对象的 parallelize()函数将列表转换为 RDD。

```
rdd = spark.sparkContext.parallelize(l)
df = spark.createDataFrame(l,['city','province','population'])
df.show()
+------------+----------+--------------+
|    city    | province | population   |
+------------+----------+--------------+
|  Nanjing   | Jiangsu  |        8.335 |
| Hangzhou   | Zhejiang |        9.468 |
|   Fuzhou   | Fujian   |         7.64 |
+------------+----------+--------------+
```

使用数据框的 columns 和 dtypes 属性得到列名和数据类型。

```
df.columns
## ['city', 'province', 'population']
df.dtypes
## [('city', 'string'), ('province', 'string'), ('population', 'double')]
```

2. 通过读入数据文件创建

调用 spark 的 read()函数得到从外部存储系统读取数据的接口，进一步级联调用其 csv()函数读取 CSV 文件，其中：

- path 表示文件路径；
- header 表示数据文件的第一行是否为变量名称，默认值为'false'；
- inferSchema 表示是否自动推测数据类型，默认值为'false'；
- nullValue 表示空值的字符串。

以下例子读取航班数据集。

```
flights = spark.read.csv('file:///root/data/flights/flights14.csv',
  header = 'true', inferSchema = 'true')
flights.show(5)
+----+------+---+----------+------------+---------+----------+----------+-------+-------+------+------+----+---------+
----------+ ----+---+
|year|month|day|dep_time|dep_delay|arr_time|arr_delay|cancelled|carrier|tailnum|flight|origin|dest|air_time|
distance|hour|min|
+----+--+--+-------+---+-------+---+--+-----+-----------+------+-----+--------+----+------+----+---+
|2014| 1 | 1 | 914   | 14 | 1238 | 13 | 0 | AA | N338AA |  1  | JFK | LAX  | 359 | 2475|  9 | 14 |
|2014| 1 | 1 | 1157  | -3 | 1523 | 13 | 0 | AA | N335AA |  3  | JFK | LAX  | 363 | 2475| 11 | 57 |
|2014| 1 | 1 | 1902  | 2  | 2224 |  9 | 0 | AA | N327AA | 21  | JFK | LAX  | 351 | 2475| 19 |  2 |
|2014| 1 | 1 | 722   |-8  | 1014 |-26 | 0 | AA | N3EHAA | 29  | LGA| PBI   | 157 | 1035|  7 | 22 |
|2014| 1 | 1 | 1347  | 2  | 1706 |  1 | 0 | AA | N319AA | 117 | JFK | LAX  | 350 | 2475| 13 | 47 |
+----+--+--+-------+---+-------+---+--+-----+-----------+------+-----+--------+----+------+----+---+
```

再读取航空公司数据集。

```
carriers = spark.read.csv('file:///root/data/flights/carrier.csv',
  header = 'true', inferSchema = 'true')
carriers.show(5)
+---------+-------------------------+
| carrier |        description       |
+---------+-------------------------+
|      02Q|Titan Airways (20...  |
|      04Q|Tradewind Aviatio...  |
|      06Q|Master Top Linhas...  |
|      07Q|Flair Airlines Lt...  |
|      09Q|Swift Air, LLC d/...  |
+---------+-------------------------+
```

10.2 数据框的选择

1. 数据列的选择

调用数据框的 select()函数选取指定列，输入参数可以是多个列名或表达式，列名的形式可以是字符串，也可以是形如<数据框名>.<列名>返回的 Column 对象。如果列名中包含*，则返回所有列，最终返回结果还是一个数据框。

以下例子选取所有列数据。

```
df = flights.select("*")
df.show(5)
+----+------+----+---------+-----------+---------+----------+----------+--------+-------+------+------+----+---------+
---------+----+---+
|year|month|day|dep_time|dep_delay|arr_time|arr_delay|cancelled|carrier|tailnum|flight|origin|dest|air_time|
distance|hour|min|
+----+--+--+------+-----+-----+-----+-----+------+-----------+-----+------+------+-----+-----+------+---+
|2014| 1 | 1 |   914|   14| 1238|   13|    0|   AA| N338AA |    1 | JFK  | LAX | 359 | 2475|    9| 14|
|2014| 1 | 1 | 1157 |   -3| 1523|   13|    0|   AA| N335AA |    3 | JFK  | LAX | 363 | 2475|   11| 57|
|2014| 1 | 1 | 1902 |    2| 2224|    9|    0|   AA| N327AA |   21 | JFK  | LAX | 351 | 2475|   19|  2|
|2014| 1 | 1 |   722|   -8| 1014|  -26|    0|   AA| N3EHAA|   29 | LGA| PBI   | 157 | 1035|    7| 22|
|2014| 1 | 1 | 1347|    2| 1706|    1|    0|   AA| N319AA | 117 | JFK  | LAX | 350 | 2475|   13| 47|
+----+--+--+------+-----+-----+-----+-----+------+-----------+-----+------+------+-----+-----+------+---+
```

以下例子通过列名的字符串选取多列数据。

```
df = flights.select('year', 'month')
df.show(5)
+----+-------+
| year|month |
+----+-------+
|2014|      1|
|2014|      1|
|2014|      1|
|2014|      1|
|2014|      1|
+----+-------+
```

以下例子通过形如<数据框名>.<列名>的 Column 对象选取多列数据。

```
df = flights.select(flights.year, flights.month)
df.show(5)
+----+-------+
|year |month|
+----+-------+
|2014|      1|
|2014|      1|
|2014|      1|
|2014|      1|
|2014|      1|
+----+------+
```

2．数据列的计算

Column 对象也可以做简单运算。以下例子通过 Column 对象表达式选取多列数据。

```
df = flights.select(flights.year - 2000, flights.month)
df.show(5)
```

```
+-----------------+-------+
|(year - 2000)  |month|
+-----------------+-------+
|              14|       1|
|              14|       1|
|              14|       1|
|              14|       1|
|              14|       1|
+-----------------+-------+
```

可以看出，以上返回数据框中的第一列的列名为(year - 2000)。

调用 Column 对象的 alias()函数设置列名。

```
df = flights.select((flights.year - 2000).alias("year2"), flights.month)
df.show(5)
+------+-------+
|year2 |month |
+------+-------+
|    14|       1|
|    14|       1|
|    14|       1|
|    14|       1|
|    14|       1|
+------+-------+
```

3. 数据行的选择

调用数据框的 filter( )函数或 where( )函数按条件筛选行，输入参数可以是 Column 对象的判断表达式，也可以是 SQL 表达式字符串。最终返回结果还是一个数据框。

以下例子通过 Column 对象的判断表达式选取航班数据集中出发地（变量 origin）为'JFK'的数据行。

```
df = flights.filter(flights.origin == 'JFK')
df.show(5)
+----+-------+---+----------+------------+--------+----------+----------+--------+--------+-----+------+----+---------+
---------+----+---+
|year|month|day|dep_time|dep_delay|arr_time|arr_delay|cancelled|carrier|tailnum|flight|origin|dest|air_time|
distance|hour|min|
+----+--+--+------+----+------+------+----+----+----------+----+----+------+----+-----+--+---+
|2014| 1 | 1 |  914|   14| 1238|    13|   0| AA| N338AA|    1| JFK| LAX| 359|2475|  9| 14|
|2014| 1 | 1 | 1157|   -3| 1523|    13|   0| AA| N335AA|    3| JFK| LAX| 363|2475|11| 57|
|2014| 1 | 1 | 1902|    2| 2224|     9|   0| AA| N327AA|   21| JFK| LAX| 351|2475|19|  2|
|2014| 1 | 1 | 1347|    2| 1706|     1|   0| AA| N319AA| 117| JFK| LAX| 350|2475|13| 47|
|2014| 1 | 1 | 2133|   -2|   37|   -18|   0| AA| N323AA| 185| JFK| LAX| 338|2475|21| 33|
+----+--+--+------+----+------+------+----+----+----------+----+----+------+----+-----+--+---+
```

如果有多个条件，可以使用运算符&表示与，|表示或。

以下例子选取航班数据集中出发地（变量 origin）为'JFK'、月份（变量 month）为 6 月的数据行。

```
df = flights.filter((flights.origin == 'JFK') & (flights.month == 6))
df.show(5)
+----+-----+----+----------+----------+---------+----------+----------+-------+-------+------+-----+----+--------+
---------+----+---+
|year|month|day|dep_time|dep_delay|arr_time|arr_delay|cancelled|carrier|tailnum|flight|origin|dest|air_time|
distance|hour|min|
+----+--+---+-----+----+----+-----+----+----+----------+--+-----+-----+-----+----+---+---+
|2014| 6 | 1 | 851  | -9  |1205| -5   |    0| AA| N787AA| 1 | JFK| LAX| 324|2475| 8 | 51 |
|2014| 6 | 1 |1220 |-10 |1522|-13  |    0| AA| N795AA| 3 | JFK| LAX| 329|2475|12| 20 |
|2014| 6 | 1 | 718  | 18 |1014| -1   |    0| AA| N784AA| 9 | JFK| LAX| 326|2475| 7 | 18 |
|2014| 6 | 1 |1024 | -6  |1314|-16  |    0| AA| N791AA|19| JFK| LAX| 320|2475|10| 24 |
|2014| 6 | 1 |1841 | -4  |2125|-45  |    0| AA| N790AA|21| JFK| LAX| 326|2475|18| 41 |
+----+--+---+-----+----+----+-----+----+----+----------+--+-----+-----+-----+----+---+---+
```

以下例子中的筛选条件为 SQL 表达式字符串。

```
df = flights.filter("origin = 'JFK' and month = 6")
df.show(5)
+----+------+----+---------+-----------+---------+----------+-----------+-------+-------+------+------+----+--------+
---------+----+---+
|year|month|day|dep_time|dep_delay|arr_time|arr_delay|cancelled|carrier|tailnum|flight|origin|dest|air_time|
distance|hour|min|
+----+--+---+-----+----+----+-----+----+----+----------+--+-----+-----+-----+----+--+---+
|2014| 6 | 1 | 851  | -9  |1205| -5   |    0| AA| N787AA| 1 | JFK | LAX| 324|2475| 8 | 51|
|2014| 6 | 1 |1220 |-10 |1522|-13  |    0| AA| N795AA| 3 | JFK | LAX| 329|2475|12| 20|
|2014| 6 | 1 | 718  | 18 |1014| -1   |    0| AA| N784AA| 9 | JFK | LAX| 326|2475| 7 | 18|
|2014| 6 | 1 |1024 | -6  |1314|-16  |    0| AA| N791AA|19| JFK | LAX| 320|2475|10| 24|
|2014| 6 | 1 |1841 | -4  |2125|-45  |    0| AA| N790AA|21| JFK | LAX| 326|2475|18| 41|
+----+--+---+-----+----+----+-----+----+----+----------+--+-----+-----+-----+----+--+---+
```

调用数据框的 limit()函数选取数据框的前几行，返回结果还是一个数据框。

```
df = flights.limit(3)
df.show()
+----+------+----+----------+-----------+---------+----------+-----------+-------+-------+-----+------+----+--------+
---------+----+---+
|year|month|day|dep_time|dep_delay|arr_time|arr_delay|cancelled|carrier|tailnum|flight|origin|dest|air_time|
distance|hour|min|
+----+--+---+-----+----+----+---+----+---+-----------+--+-----+-----+----+----+--+---+
|2014| 1 | 1 | 914 | 14  |1238| 13|    0| AA| N338AA| 1 | JFK| LAX| 359|2475| 9 | 14|
|2014| 1 | 1 |1157 | -3  |1523| 13|    0| AA| N335AA| 3 | JFK| LAX| 363|2475|11| 57|
|2014| 1 | 1 |1902 | 2   |2224|  9|    0| AA| N327AA|21| JFK| LAX| 351|2475|19|  2 |
+----+--+---+-----+----+----+---+----+---+-----------+--+-----+-----+----+----+--+---+
```

10.3 数据框的运算和聚合

1. 数据框的统计运算

程序包 pyspark.sql.functions 中有如下函数实现常用的统计功能，输入参数可以是多个列名或表达式，列名的形式可以是字符串，也可以是形如<数据框名>.<列名>返回的 Column 对象，如表 10-1 所示。

表 10-1 数据框的常用统计聚合函数

函　　数	功　　能
count()	非空元素个数
countDistinct()	不重复元素个数
max()和 min()	最大和最小的元素
avg()和 mean()	均值
variance()	方差
stddev()	标准差
sum()	总和
skewness()	偏度
kurtosis()	峰度

```
flights.select(count('arr_delay'), count('dep_delay')).show()
+----------------+------------------+
|count(arr_delay)|count(dep_delay)|
+----------------+------------------+
|          253316|           253316|
+----------------+------------------+
flights.select(countDistinct('arr_delay'), countDistinct('dep_delay')).show()
+----------------+------------------+
|count(arr_delay)|count(dep_delay)|
+----------------+------------------+
|             616|              570|
+----------------+------------------+
flights.select(max('arr_delay'), max('dep_delay')).show()
+----------------+------------------+
|max(arr_delay) |max(dep_delay)   |
+----------------+------------------+
|            1494|             1498|
+----------------+------------------+
flights.select(mean('arr_delay'), mean('dep_delay')).show()
+------------------+---------------------+
| avg(arr_delay)  |    avg(dep_delay)   |
+------------------+---------------------+
|8.14670214277819|12.465264728639328|
+------------------+---------------------+
flights.select(variance('arr_delay'), variance('dep_delay')).show()
+-------------------------+--------------------------+
|var_samp(arr_delay,0,0) |var_samp(dep_delay,0,0)|
+-------------------------+--------------------------+
|    2093.012313739685 |    1734.6194791636824|
+-------------------------+--------------------------+
flights.select(stddev('arr_delay'), stddev('dep_delay')).show()
+-----------------------------+------------------------------+
```

```
|stddev_samp(arr_delay,0,0) |stddev_samp(dep_delay,0,0)|
+-----------------------------+-------------------------------+
|         45.74945151299286 |          41.64876323690395 |
+-----------------------------+-------------------------------+
flights.select(skewness('arr_delay'), skewness('dep_delay')).show()
+-----------------------------+-------------------------------+
|  skewness(arr_delay,0,0)   |  skewness(dep_delay,0,0)   |
+-----------------------------+-------------------------------+
|         4.696522514398242 |          5.846543930742305 |
+-----------------------------+-------------------------------+
flights.select(kurtosis('arr_delay'), kurtosis('dep_delay')).show()
+-----------------------------+-------------------------------+
|  kurtosis(arr_delay,0,0)   |  kurtosis(dep_delay,0,0)   |
+-----------------------------+-------------------------------+
|         48.81973252661658|          68.19506524696821|
+-----------------------------+-------------------------------+
```

同样的，这些函数参数可以是形如<数据框名>.<列名>的 Column 对象，并可以调用 alias()函数设置列名。

```
flights.select(mean(flights.arr_delay).alias("arr_delay_mean"),
mean(flights.dep_delay).alias("dep_delay_mean")).show()
+--------------------+----------------------+
|    arr_delay_mean|       dep_delay_mean|
+--------------------+----------------------+
|8.14670214277819|12.465264728639328|
+--------------------+----------------------+
```

2. 数据框的聚合

数据聚合通常可以分为以下 3 个步骤：

- 根据某些条件将数据记录分成不同组；
- 对每个组独立应用某个函数；
- 将不同组得到的结果合并成某种数据结构。

调用数据框的 groupby()函数做数据分组，输入参数可以是多个列名或表达式，返回一个 GroupedData 分组对象。

以下例子按出发地（变量 origin）分组。

```
grouped = flights.groupby('origin')
print(grouped)
## <pyspark.sql.group.GroupedData object at 0x7f9c3016bf10>
```

调用 GroupedData 分组对象的 count()函数返回包含分组记录数的数据框。

```
grouped.count().show()
+-------+-------+
| origin | count |
+-------+-------+
| EWR | 87400 |
```

```
| LGA | 84433|
|  JFK | 81483|
+-------+-------+
```

GroupedData 分组对象有如下函数实现常用统计功能，如表 10-2 所示。

表 10-2　GroupedData 分组对象的常用统计聚合函数

函　　数	功　　能
count()	非空元素个数
max()和 min()	最大和最小的元素
avg()和 mean()	均值
sum()	总和

```
grouped.max('arr_delay','dep_delay').show( )
+----------+----------------+-----------------+
|  origin  |max(arr_delay)|max(dep_delay)|
+----------+----------------+-----------------+
|  EWR   |          1494 |           1498|
|  LGA   |           996 |            973|
|   JFK   |          1223 |           1241|
+----------+----------------+-----------------+
grouped.mean('arr_delay','dep_delay').show( )
+----------+----------------------+-----------------------+
|  origin  |  avg(arr_delay)       |   avg(dep_delay)      |
+----------+----------------------+-----------------------+
| EWR      |10.026121281464532 |15.212482837528604 |
| LGA      |  6.60196842466808 |10.605000414529863 |
| JFK      | 7.731465459052808 |11.446166685075415 |
+----------+----------------------+-----------------------+
grouped.sum('arr_delay','dep_delay').show( )
+----------+----------------+----------------+
|origin    |sum(arr_delay) |sum(dep_delay)|
+----------+----------------+----------------+
|   EWR  |        876283 |       1329571 |
|   LGA  |        557424 |        895412 |
|    JFK  |        629983 |        932668 |
+----------+----------------+----------------+
```

调用 GroupedData 分组对象的 agg()函数可以实现更高级的统计功能，输入参数可以是键为变量名、值为聚合函数的字典，也可以是由程序包 pyspark.sql.functions 中统计函数得到的 Column 对象列表。

以下例子在 agg( )函数中指定由程序包 pyspark.sql.functions 中统计函数得到的 Column 对象列表，计算在各不同出发地和目的地组合中，平均到达和出发延误分钟数（变量 arr_delay 和 dep_delay）。

```
flights.groupby('origin', 'dest').agg(mean('arr_delay'), mean('dep_delay')).show(5)
+----------+-----------+----------------------+-----------------------+
|  origin  |  dest    |  avg(arr_delay)       |  avg(dep_delay)       |
```

```
+-----------+-----------+----------------------+-----------------------+
|  LGA   | BTV      | 7.637554585152839 | 17.88646288209607 |
|  LGA   | ROA      |             6.672 |             6.368 |
|  EWR   | TUL      |27.572139303482587|28.895522388059703|
|  JFK   | OAK      |14.417004048582996|15.898785425101215|
|  LGA   | ROC      |11.740157480314961| 22.04724409448819 |
+-----------+-----------+----------------------+-----------------------+
```

以下例子在函数 agg()中指定键为变量名、值为聚合函数的字典，即对不同变量执行不同的聚合函数，计算在各不同出发地和目的地组合中，到达延误分钟数（变量 arr_delay）的均值和出发延误分钟数（变量 dep_delay）的最大值。

```
flights.groupby('origin', 'dest').agg({'arr_delay':'mean', 'dep_delay':'max'}).show(5)
+-----------+-----------+----------------------+-------------------+
|  origin  |  dest   |    avg(arr_delay)   |max(dep_delay) |
+-----------+-----------+----------------------+-------------------+
|   LGA   | BTV      | 7.637554585152839 |           319  |
|   LGA   | ROA      |             6.672|           134  |
|   EWR   | TUL      |27.572139303482587|           255  |
|   JFK   | OAK      |14.417004048582996|           320  |
|   LGA   | ROC      |11.740157480314961|           304  |
+-----------+-----------+----------------------+-------------------+
```

10.4 数据框的增加、删除和修改

1. 数据增加

调用数据框的函数 withColumn()增加或替换数据列，返回一个新的数据框，其中

- 第 1 个参数 colName 表示增加或替换的数据列的名称；
- 第 2 个参数 col 表示 Column 对象的表达式。

以下例子增加 2 列：平均速度（变量 speed）和总延误分钟数（变量 delay）。

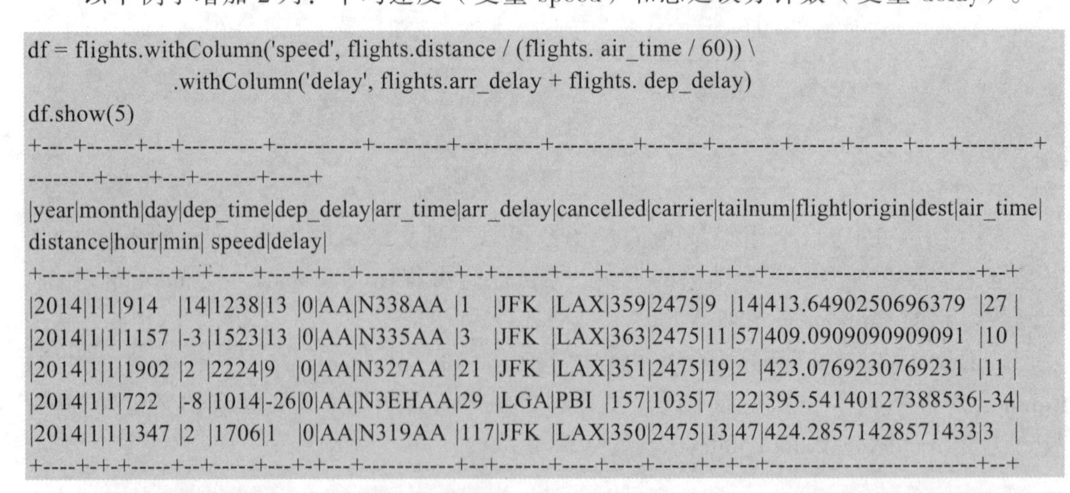

```
df = flights.withColumn('speed', flights.distance / (flights. air_time / 60)) \
              .withColumn('delay', flights.arr_delay + flights. dep_delay)
df.show(5)
+----+------+---+---------+----------+--------+---------+---------+-------+--------+------+------+----+---------+
--------+-----+---+-------+-----+
|year|month|day|dep_time|dep_delay|arr_time|arr_delay|cancelled|carrier|tailnum|flight|origin|dest|air_time|
distance|hour|min| speed|delay|
+----+-+-+-----+--+-----+---+-+---+----------+--+------+----+----+-----+--+--+----------------------+--+
|2014|1|1|914   |14|1238|13 |0|AA|N338AA |1   |JFK |LAX|359|2475|9  |14|413.6490250696379  |27 |
|2014|1|1|1157 |-3 |1523|13 |0|AA|N335AA |3   |JFK |LAX|363|2475|11 |57|409.0909090909091  |10 |
|2014|1|1|1902 |2  |2224|9  |0|AA|N327AA |21  |JFK |LAX|351|2475|19|2  |423.0769230769231  |11 |
|2014|1|1|722   |-8 |1014|-26|0|AA|N3EHAA|29  |LGA|PBI  |157|1035|7  |22|395.54140127388536|-34|
|2014|1|1|1347 |2  |1706|1  |0|AA|N319AA |117|JFK |LAX|350|2475|13|47|424.28571428571433|3   |
+----+-+-+-----+--+-----+---+-+---+----------+--+------+----+----+-----+--+--+----------------------+--+
```

调用数据框的 unionAll()函数与另一个数据框按行合并，返回一个新的数据框。以下例子将航班数据的前 3 行与航班数据所有行按行合并。

```
df = flights.limit(3).unionAll(flights)
df.show(5)
+----+-----+----+----------+-----------+---------+----------+----------+-------+--------+------+------+----+---------+
---------+----+---+
|year|month|day|dep_time|dep_delay|arr_time|arr_delay|cancelled|carrier|tailnum|flight|origin|dest|air_time|
distance|hour|min|
+----+--+--+------+----+----+---+-+---+----------+--+----+------+----+------+----+-----+
|2014| 1 | 1 |914  |14  |1238|13 |0 |AA|N338AA|1 |JFK |LAX |359 |2475 |9   |14  |
|2014| 1 | 1 |1157 |-3  |1523|13 |0 |AA|N335AA|3 |JFK |LAX |363 |2475 |11  |57  |
|2014| 1 | 1 |1902 |2   |2224|9  |0 |AA|N327AA|21|JFK |LAX |351 |2475 |19  |2   |
|2014| 1 | 1 |914  |14  |1238|13 |0 |AA|N338AA|1 |JFK |LAX |359 |2475 |9   |14  |
|2014| 1 | 1 |1157 |-3  |1523|13 |0 |AA|N335AA|3 |JFK |LAX |363 |2475 |11  |57  |
+----+--+--+------+----+----+---+-+---+----------+--+----+------+----+------+----+-----+
```

2．数据删除

调用数据框的 drop()函数删除指定列，返回一个新的数据框。以下例子删除航班数据的变量 tailnum。

```
df = flights.drop('tailnum')
df.show(5)
+----+------+---+----------+-----------+----------+----------+---------+-------+------+------+----+--------+---------+
----+---+
|year|month|day|dep_time|dep_delay|arr_time|arr_delay|cancelled|carrier|flight|origin|dest|air_time|distance|
hour|min|
+----+--+--+-----+-----+-----+----+--+---+---+-----+------+---+-----+---+---+
|2014| 1 | 1 |914  |14  |1238 |13  |0 |AA |1   |JFK |LAX|359 |2475|9  |14 |
|2014| 1 | 1 |1157 |-3  |1523 |13  |0 |AA |3   |JFK |LAX|363 |2475|11 |57 |
|2014| 1 | 1 |1902 |2   |2224 |9   |0 |AA |21  |JFK |LAX|351 |2475|19 |2  |
|2014| 1 | 1 |722  |-8  |1014 |-26 |0 |AA |29  |LGA|PBI |157 |1035|7  |22 |
|2014| 1 | 1 |1347 |2   |1706 |1   |0 |AA |117|JFK |LAX|350 |2475|13 |47 |
+----+--+--+-----+-----+-----+----+--+---+---+-----+------+---+-----+---+---+
```

3．数据修改

调用数据框的 withColumn()函数增加或替换数据列，返回一个新的数据框，其中：

- 第 1 个参数 colName 表示增加或替换的数据列的名称；
- 第 2 个参数 col 表示 Column 对象的表达式。

调用数据框的 withColumnRenamed()函数修改列名，返回一个新的数据框，其中：

- 第 1 个参数 existing 表示修改前的列名；
- 第 2 个参数 new 表示修改后的列名。

```
df = flights.withColumnRenamed('cancelled','cancelled_flag')
df.show(5)
+----+-------+----+----------+------------+----------+-----------+-----------------+-------+-------+------+------+----+
----------+--------+-----+---+
|year|month|day|dep_time|dep_delay|arr_time|arr_delay|cancelled_flag|carrier|tailnum|flight|origin|dest|
air_time|distance|hour|min|
```

```
+----+---+--+------+----+------+-----+-+----+-----------+----+------+-----+----+------+---+---+
|2014| 1 | 1 |914   |14  |1238 |13   |0 |AA |N338AA  | 1   |JFK |LAX|359 |2475 |9   |14 |
|2014| 1 | 1 |1157 |-3   |1523 |13   |0 |AA |N335AA  |3    |JFK |LAX|363 |2475 |11 |57 |
|2014| 1 | 1 |1902 |2    |2224 |9    |0 |AA |N327AA  |21   |JFK |LAX|351 |2475 |19 |2  |
|2014| 1 | 1 |722  |-8   |1014 |-26 |0 |AA |N3EHAA |29  |LGA|PBI |157 |1035 |7   |22 |
|2014| 1 | 1 |1347 |2    |1706 |1    |0 |AA |N319AA  |117|JFK |LAX|350 |2475 |13 |47 |
+----+---+--+------+----+------+-----+-+----+-----------+----+------+-----+----+------+---+---+
```

10.5 数据框的连接

数据框连接是指两个数据框按指定的连接条件按列合并成一个数据框，如图 10-1 所示。主要分为以下几种：

- 内连接（inner join）：包含左侧数据框和右侧数据框中都存在且满足连接条件的行；
- 左连接（left join）：包含左侧数据框中所有行，右侧数据框中不满足连接条件的列设为空；
- 右连接（right join）：包含右侧数据框中所有行，左侧数据框中不满足连接条件的列设为空；
- 全连接（full join）：包含左侧数据框和右侧数据框中所有行，不满足连接条件的列设为空。

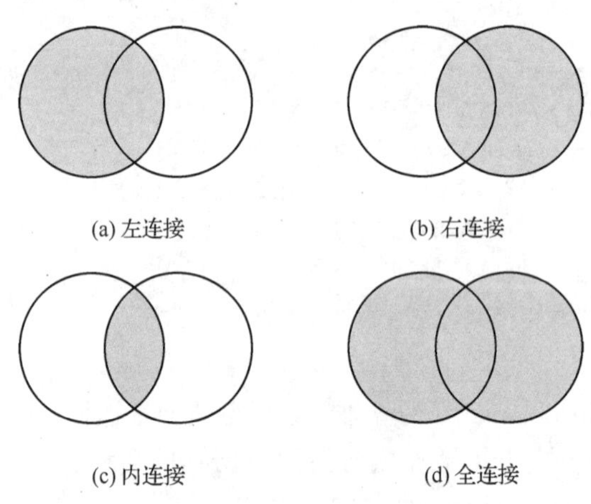

图 10-1　数据框连接分类图

数据框的 join()函数用于做数据连接，其中：

- 自身对象表示左侧数据框；
- 第 1 个参数 other 表示右侧数据框；
- 第 2 个参数 on 表示左侧和右侧数据框的连接变量名（列表）或连接条件；
- 第 3 个参数 how 表示连接方式，left_outer 为左连接，right_outer 为右连接，outer 为外连接，inner 为内连接，默认为内连接。

航班数据集（数据表 flights）中的变量 carrier 与航空公司代码描述数据集（数据表 carriers）中的变量 carrier 都表示航空公司代码，将作为连接条件。需要注意的是，航空公司代码描述数据集（数据表 carriers）中每个代码仅对应一条记录。

1. 内连接

调用数据框的 join()函数并指定参数 how = 'inner'做内连接、参数 on 为连接变量名。

```
df = flights.join(carriers, 'carrier', 'inner')
df.show(5)
+-------+---+------+----+---------+----------+---------+----------+-----------+-------+------+------+----+---------+
---------+----+---+-------------+
|carrier|year|month|day|dep_time|dep_delay|arr_time|arr_delay|cancelled|tailnum|flight|origin|dest|air_time|
distance|hour|min|description|
+---+----+--+-+-------+---+----+---+-+-----------+---+------+----+---+-----+----+-----+----------------------+
|AA|2014| 1| 1| 914    |14 |1238|13 |0 |N338AA  |1    |JFK  |LAX|359|2475|9    |14    |American Airlines...|
|AA|2014| 1| 1| 1157   |-3 |1523|13 |0 |N335AA  |3    |JFK  |LAX|363|2475|11   |57    |American Airlines...|
|AA|2014| 1| 1| 1902   |2  |2224|9  |0 |N327AA  |21   |JFK  |LAX|351|2475|19   |2     |American Airlines...|
|AA|2014| 1| 1| 722    |-8 |1014|-26|0 |N3EHAA  |29   |LGA|PBI  |157|1035|7    |22    |American Airlines...|
|AA|2014| 1| 1| 1347   |2  |1706|1  |0 |N319AA  |117|JFK  |LAX|350|2475|13   |47    |American Airlines...|
+---+----+--+-+-------+---+----+---+-+-----------+---+------+----+---+-----+----+-----+----------------------+
```

计算内连接前航班数据集的记录条数和连接后的记录条数。

```
flights.count()
## 253316
df.count()
## 235315
```

可以看出，连接后的记录条数少于连接前航班数据集的记录条数，意味着航班数据集中有一些记录并没有出现在结果中。

指定参数 on 为连接条件做内连接，与之前得到的结果完全一致，除了连接变量 carrier 出现了 2 次。

```
df = flights.join(carriers, flights.carrier == carriers.carrier, 'inner')
df.show(5)
+----+------+---+----------+-----------+----------+----------+---------+-------+--------+------+------+----+---------+
--------+-----+---+-------+------------+
|year|month|day|dep_time|dep_delay|arr_time|arr_delay|cancelled|carrier|tailnum|flight|origin|dest|air_time|
distance|hour|min|carrier|description|
+----+--+--+-----+--+----+---+-+---+----------+---+------+-----+----+-----+--+---+----+----------------------+
|2014| 1 | 1 |914  |14|1238|13 |0|AA|N338AA |1   |JFK   |LAX|359|2475|9  |14 |AA|American Airlines...|
|2014| 1 | 1 |1157|-3 |1523|13 |0|AA|N335AA |3   |JFK   |LAX|363|2475|11 |57 |AA|American Airlines...|
|2014| 1 | 1 |1902|2  |2224|9  |0|AA|N327AA |21  |JFK   |LAX|351|2475|19 |2  |AA|American Airlines...|
|2014| 1 | 1 |722  |-8|1014|-26|0|AA|N3EHAA|29  |LGA   |PBI |157|1035|7  |22 |AA|American Airlines...|
|2014| 1 | 1 |1347|2  |1706|1  |0|AA|N319AA |117|JFK   |LAX|350|2475|13 |47 |AA|American Airlines...|
+----+--+--+-----+--+----+---+-+---+----------+---+------+-----+----+-----+--+---+----+----------------------+
```

2. 左连接

调用数据框的 join()函数并指定参数 how = 'left_outer'做左连接。

```
df = flights.join(carriers, 'carrier', 'left_outer')
df.show(5)
```

```
+-------+---+-------+----+---------+-----------+---------+----------+----------+--------+------+------+----+---------+
---------+----+---+------------+
|carrier|year|month|day|dep_time|dep_delay|arr_time|arr_delay|cancelled|tailnum|flight|origin|dest|air_time|
distance|hour|min|description|
+---+----+---+--+-----+-----+-----+-----+-+----------+----+-----+-----+---+-----+--+---+----------------------+
|AA|2014| 1 | 1 |914   |14    |1238|13    |0|N338AA |1    |JFK |LAX|359|2475|9  |14 |American Airlines...|
|AA|2014| 1 | 1 |1157  |-3    |1523|13    |0|N335AA |3    |JFK |LAX|363|2475|11|57 |American Airlines...|
|AA|2014| 1 | 1 |1902  |2     |2224|9     |0|N327AA |21   |JFK |LAX|351|2475|19|2  |American Airlines...|
|AA|2014| 1 | 1 |722   |-8    |1014|-26   |0|N3EHAA|29   |LGA|PBI  |157|1035|7  |22 |American Airlines...|
|AA|2014| 1 | 1 |1347  |2     |1706|1     |0|N319AA |117 |JFK |LAX|350|2475|13|47 |American Airlines...|
+---+----+---+--+-----+-----+-----+-----+-+----------+----+-----+-----+---+-----+--+---+----------------------+
```

计算左连接前航班数据集的记录条数和连接后的记录条数。

```
flights.count( )
## 253316
df.count( )
## 235316
```

可以看出，连接后的记录条数完全等于连接前航班数据集的记录条数，意味着航班数据集中有全部记录都出现在结果中。

右连接与左连接类似，在此不再赘述。

3. 全链接

调用数据框的 join()函数并指定参数 how 为 outer 做全连接。

```
df = flights.join(carriers, 'carrier', 'outer')
df.show(5)
+------+----+------+----+---------+-----------+---------+----------+----------+--------+------+------+----+
---------+---------+-----+----+------------+
|carrier|year|month|day|dep_time|dep_delay|arr_time|arr_delay|cancelled|tailnum|flight|origin|dest|
air_time|distance|hour| min |description|
+----+----+----+---+----+---+----+----+---+----+---+----+----+---+----+---+----+-------------------------+
|AJQ |null|null|null|null|null|null|null|null|null|null|null|null|null|null|null|Aerodynamics Inc....    |
| MS  |null|null|null|null|null|null|null|null|null|null|null|null|null|null|null|Egyptair (1985 - )      |
| TB  |null|null|null|null|null|null|null|null|null|null|null|null|null|null|null|TUI Airlines Belg...    |
|0HQ |null|null|null|null|null|null|null|null|null|null|null|null|null|null|null|Polar Airlines de...    |
|CSQ |null|null|null|null|null|null|null|null|null|null|null|null|null|null|null|Celtic Tech Jet L...    |
+----+----+----+---+----+---+----+----+---+----+---+----+----+---+----+---+----+-------------------------+
```

计算全连接前航班数据集的记录条数和连接后的记录条数。

```
flights.count( )
## 253316
flights.join(carriers, flights.carrier == carriers.carrier, 'outer').count( )
## 253728
```

可以看出，连接后的记录条数大于连接前航班数据集的记录条数，意味着航空公司代码数据集（数据框 carriers）中的变量 carrier 的有些数值没有出现在航班数据集（数据框 flights）中。

10.6 数据框的变形

数据变形指的是从“长”格式转换为“宽”格式或相反。对于同一个数据集，“长”格式的表现形式如表 10-3 所示。

表 10-3 “长”格式的表现形式

维　　度	变　量　名	数　　值
A	变量 1	1
A	变量 2	2
A	变量 3	3
B	变量 1	4
B	变量 3	5

对应“宽”格式的表现形式如表 10-4 所示。

表 10-4 “宽”格式的表现形式

维　　度	变量 1	变量 2	变量 3
A	1	2	3
B	4	NA	5

调用 GroupedData 对象的 pivot()函数在一个列的各种取值上做分组统计，并以不同列的形式呈现。其中：

- 第 1 个参数 pivot_col 表示做分组统计的列名；
- 第 2 个参数 values 表示该列的可能取值，为空则自动推断，效率较低。

计算在各不同年份、月份和出发地组合中，平均出发延误分钟数(变量 dep_delay)，其中不同的出发地用不同列表示。

```
df = flights.groupBy('year', 'month').pivot('origin', ['EWR', 'JFK', 'LGA']).mean('dep_delay')
df.show()
+----+-----+------------------------+------------------------+-----------------------+
|year|month|                    EWR |                    JFK |                   LGA |
+----+-----+------------------------+------------------------+-----------------------+
|2014|    1 |24.203855068927655 | 26.98282899366643  | 17.77822257806245   |
|2014|    2 | 19.87146814404432  |19.152234842897183 |14.289991194599354  |
|2014|    3 | 11.24013504958852  | 7.171331636980492  |  8.07248044178555  |
|2014|    4 |16.478549005929544 | 6.997397769516729  |  7.16597510373444  |
|2014|    5 |17.129032258064516 |10.932931968693557 |12.934422457432122 |
|2014|    6 |18.559418119584983 | 9.807884113037284  |13.418263163932545 |
|2014|    7 |19.314775613886535 |17.563115693012602 |12.321813031161474 |
|2014|    8 |11.933425209371356 | 8.660026708212776  |  9.35167755508803  |
|2014|    9 | 6.513807629300257  |3.7615459406903256 | 3.903754752851711  |
|2014|   10 | 8.040915959548995  | 6.969552585705985  | 8.52325976230899 8 |
+----+-----+------------------------+------------------------+-----------------------+
```

以下例子没有指定参数 values，效率较低。

```
df = flights.groupBy('year', 'month').pivot('origin').mean('dep_delay')
df.show( )
+----+------+------------------------+------------------------+------------------------+
|year|month|                 EWR |                  JFK |                  LGA |
+----+------+------------------------+------------------------+------------------------+
|2014|     1 |24.203855068927655 | 26.98282899366643  | 17.77822257806245  |
|2014|     2 | 19.87146814404432 |19.152234842897183 |14.289991194599354 |
|2014|     3 | 11.24013504958852 | 7.171331636980492 |  8.07248044178555 |
|2014|     4 |16.478549005929544 | 6.997397769516729 |  7.16597510373444 |
|2014|     5 |17.129032258064516 |10.932931968693557 |12.934422457432122 |
|2014|     6 |18.559418119584983 | 9.807884113037284 |13.418263163932545 |
|2014|     7 |19.314775613886535 |17.563115693012602 |12.321813031161474 |
|2014|     8 |11.933425209371356 | 8.660026708212776 |  9.35167755508803 |
|2014|     9 | 6.513807629300257 |3.7615459406903256 | 3.903754752851711 |
|2014|    10 | 8.040915959548995 | 6.969552585705985 | 8.523259762308998 |
+----+------+------------------------+------------------------+------------------------+
```

小　　结

本章主要介绍了 Spark 大数据处理的基本操作，包括数据框的创建、选择、聚合、引用、快速筛选、连接和变形。数据框在理念上类似于关系型数据库中的数据表以及 R 语言和 Python 中的数据框，但在性能上更加优化。

本章使用 Spark 中的 Spark SQL 实现大数据处理的各个环节。Spark SQL 的 API 提供更多关于数据结构和执行计算的信息。Spark SQL 的执行引擎会利用这些额外信息作系统优化。

习　　题

1. 数据框与 RDD 相比的主要特点是什么？
2. 内连接、左连接和右连接有什么不同？
3. 基于航班准点数据集，哪个月份的平均到达延误分钟数最大？
4. 基于航班准点数据集，哪个航空公司（名称）的飞机数量（通过不同的尾翼编号确定）最多？

Spark 机器学习流程 ‹‹‹ 第 11 章

MLlib 是 Spark 中用于机器学习的模块，其宗旨是使用户能够以简便和可扩展的方式做机器学习。MLlib 中包含了许多常用机器学习算法，包括分类、回归、聚类、协同过滤和降维等。MLlib 又进一步分为两个程序包：

- spark.mllib 程序包中的 API 针对于 RDD 数据集；
- spark.ml 程序包中的 API 针对于数据框。

spark.ml 程序包是 Spark 主推的机器学习包，而 spark.mllib 程序包自从 Spark 2.0 开始就已进入维护状态，不再添加新的功能。

本章主要介绍使用 MLlib 做机器学习。该程序包可以用于数据的预处理、后处理、特征工程、模型训练、超参数调优、模型测试和模型评估，用户只需要简单地调用 MLlib 里的模块即可实现这些功能，并不需要具体实现所有的算法。

本章所有例子基于一个泰坦尼克号乘客数据集，预测泰坦尼克号乘客的命运（是否生存），包含了将近 80%乘客的信息和生存状态，包含 1 309 个样本，每个样本包含 14 个属性，如表 11-1 所示

表 11-1　泰坦尼克号乘客数据集属性定义

属　　性	定　　义
pclass	舱位（“1st”为一等舱，“2nd”为二等舱，“3rd”为三等舱）
survived	是否生存（1 为生存，0 为未生存）
name	乘客姓名
sex	性别
age	年龄
sibsp	在船上的配偶和兄弟姐妹数量
parch	在船上的父母和子女数量
ticket	船票号码
fare	票价
cabin	房间号
embarked	登船地点
boat	救生船号码
body	尸体编号
home.dest	家乡

该数据集的路径为～/data/flights/flights14.csv。

使用 pyspark 命令启动 Python 版本的 Spark Shell。

```
pyspark
```

载入本章需要用到的程序包。

```
from pyspark.sql.functions import mean, isnan, isnull
from pyspark.ml.feature import VectorAssembler, StringIndexer, OneHotEncoderEstimator, Imputer,
ChiSqSelector
from pyspark.ml.tuning import ParamGridBuilder, CrossValidator
from pyspark.ml.classification import DecisionTreeClassifier
from pyspark.ml.evaluation import BinaryClassificationEvaluator, MulticlassClassificationEvaluator
```

调用 spark 的 read()函数读取泰坦尼克号乘客数据集。

```
titanic3 = spark.read.csv('file:///root/data/titanic/titanic3.csv',
   header ='true', inferSchema ='true', nullValue = 'NA')
titanic3.show(5)
+---------+-----------+--------+----+-----+----+------+-------+------+-------+------------+-----+----+-------------+
| pclass | survived | name| sex| age|sibsp| parch| ticket| fare| cabin| embarked| boat|body| home.dest|
+-+-+--------------------+-----+------------+++-------+----------+--------+------------+-+---+--------------------+
|1st|1|Allen,Miss.Elis.. |female| 29.0 |0|0|24160 |211.337494| B5 |Southampton| 2 |null| St Louis, MO |
|1st|1|Allison,Master.... |male |0.916700006|1|2|113781|151.550003|C22 C26|Southampton|11|null|Montreal, PQ/Ch... |
|1st|0|Allison, Miss. He... |female| 2.0 |1|2|113781|151.550003|C22 C26|Southampton| |null|Montreal, PQ / Ch... |
|1st|0|Allison, Mr. Huds... | male | 30.0 |1|2|113781|151.550003|C22 C26|Southampton| |135|Montreal, PQ / Ch... |
|1st|0|Allison, Mrs. Hud... |female| 25.0 |1|2|113781|151.550003|C22 C26|Southampton| |null|Montreal, PQ / Ch... |
+-+-+--------------------+-----+------------+++-------+----------+--------+------------+-+---+--------------------+
```

11.1 数据探索

在做进一步数据建模前，首先需要对数据变量名、数值分布和缺失值情况等等有初步了解。

调用数据框的函数 describe( )得到数据框中各数值型列的统计信息，包括数据集中变量名、非缺失值的数量（count）、均值（mean）、标准差（stddev）、最小值（min）和最大值（max）。如果没有指定任何列，这个函数统计所有数值列。

调用数据框的函数 count()得到数据框的样本数。

```
titanic3.describe( ).show( )
+----------------+-----------------+---------------+-------------+--------------+-------------+----------------+
| summary | survived | age | sibsp | parch | fare | body |
+----------------+-----------------+---------------+-------------+--------------+-------------+----------------+
| count| 1309 | 1046 | 1309| 1309| 1308| 121|
| mean |0.3819709702062643|29.881134512434034|0.4988540870893812|0.3850267379679144|
33.29547938641437|160.8099173553719|
| stddev|0.4860551708664828|14.413499699911787|1.0416583905961025|0.8655602753495145|
51.75866882693879|97.69692199600308|
| min| 0| 0.166700006| 0| 0| 0.0| 1|
| max| 1| 80.0| 8| 9| 512.329224| 328|
+----------------+-----------------+---------------+-------------+--------------+-------------+----------------+
titanic3.count( )
## 1309
```

可以看出，3 个变量 age、fare 和 body 存在缺失值；变量 fare 的最大和最小值差异较大。

11.2 数据划分

通常把分类错误的样本数占样本总数的比例称为错误率（error rate），即如果在 m 个样本中有 a 个样本分类错误，则错误率 $E=a/m$；相应的，$1-a/m$ 称为精度（accuracy）。模型在训练集上的误差称为训练误差，在新样本上的误差称为泛化误差（generalization error）。

一般希望得到泛化误差小的模型，应该从训练样本中尽可能学出适用于所有潜在样本的普遍规律。然而，当模型把训练样本学得太好时，很可能已经把训练样本自身的一些特点当作所有潜在样本都会具有的一般性质，导致泛化性能下降，这种现象称为过拟合（overfitting）。与过拟合相对的是欠拟合，是指对训练样本的一般性质尚未学好。图 11-1 给出的例子便于直观理解。

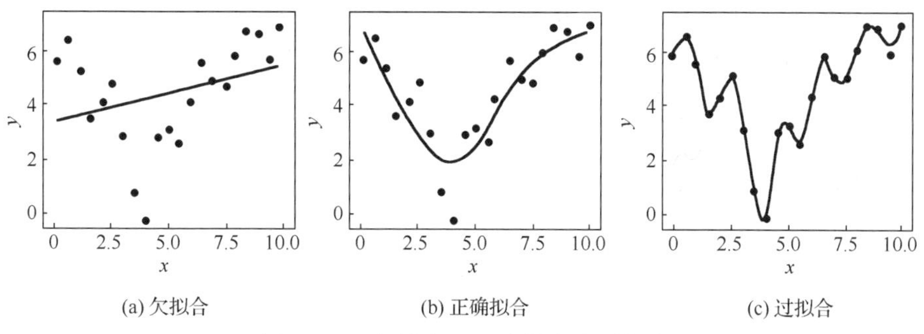

图 11-1 欠拟合、正确拟合和过拟合

通常，人们通过实验测试来对模型的泛化误差进行评估，为此需要使用一个测试集来测试模型对新样本的判别能力，然后以测试集上的测试误差作为泛化误差的近似。而该测试集通常仅能使用一次，如果多次使用测试集做评估，则无形之中测试集也充当了训练模型的作用，则测试误差会低估泛化误差。因此，我们会进一步将训练集划分成训练集和验证集，使用验证集多次模型进行评估并进而做出选择，如图 11-2 所示。

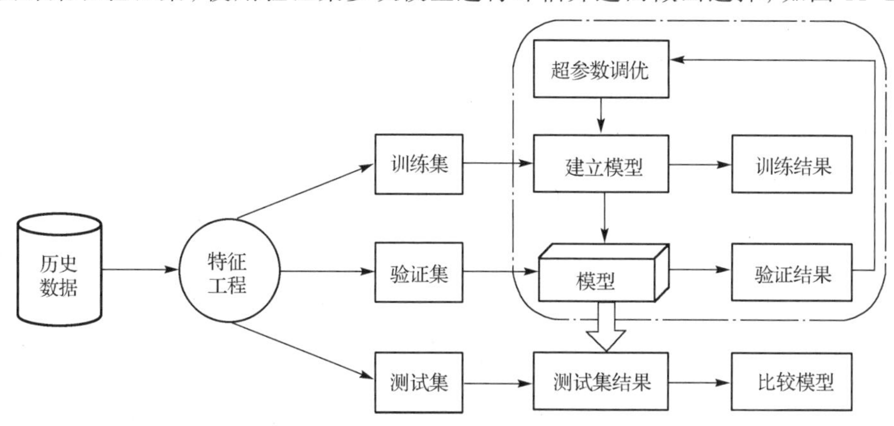

图 11-2 训练集、验证集和测试集

调用数据框的 randomSplit()函数将数据集随机划分为训练集和测试集，其中：

- 第 1 个参数 weights 表示分隔后各部分的比例。
- 第 2 个参数 seed 表示随机数的种子。

这里随机抽取 70%的样本作为训练集，并且仅保留有预测意义的变量。

```
titanic3_sub = titanic3.select(['pclass','sex','age','sibsp', 'parch','fare','embarked','survived'])
train, test = titanic3_sub.randomSplit([0.7, 0.3], 123)
train.count( )
## 910
test.count( )
## 399
```

11.3 数据填充

在前几节中，可以看出变量 age、fare 和 body 存在缺失值。有些模型可以接受有缺失值的数据样本，而有些模型要求数据样本完全没有缺失值。

调用 pyspark.sql.functions 程序包中的 isnull()函数得到变量 age 和 fare 是否为空的布尔型列。

```
age_train_na = isnull(train['age'])
age_test_na = isnull(test['age'])
fare_train_na = isnull(train['fare'])
```

1. 填充训练集

调用 pyspark.ml.feature 程序包中的 Imputer()构造函数创建填充器，该函数主要是对浮点型列进行操作，其中：

- 第 1 个参数 strategy 表示填充策略，mean 为用均值填充，median 为用中位数填充，默认为 mean；
- 第 2 个参数 missing_values 表示缺失值的表示方式，默认为 nan；
- 第 3 个参数 inputCols 表示输入变量名列表；
- 第 4 个参数 outputCols 表示输出变量名列表；
- 返回结果为一个填充器对象。

以下例子用中位数填充所有训练集中的数值变量，即变量 age 和 fare。

调用填充器的 fit()函数做训练，输入参数为数据框，返回训练后的填充器模型。调用填充器模型的 transform()函数填充训练集，输入参数和返回结果都是数据框。

```
imp = Imputer(strategy='median', inputCols=['age', 'fare'], outputCols=['age', 'fare'])
impm = imp.fit(train)
train = impm.transform(train)
```

查看在原数据集中变量 age 和 fare 为空的记录填充后的情况。

```
train.where(age_train_na).show(5)
+-------+--------+----+-------+-------+--------------+---------------+------------+
|pclass |   sex  | age|sibsp  |parch  |     fare     |  embarked     |survived    |
+-------+--------+----+-------+-------+--------------+---------------+------------+
```

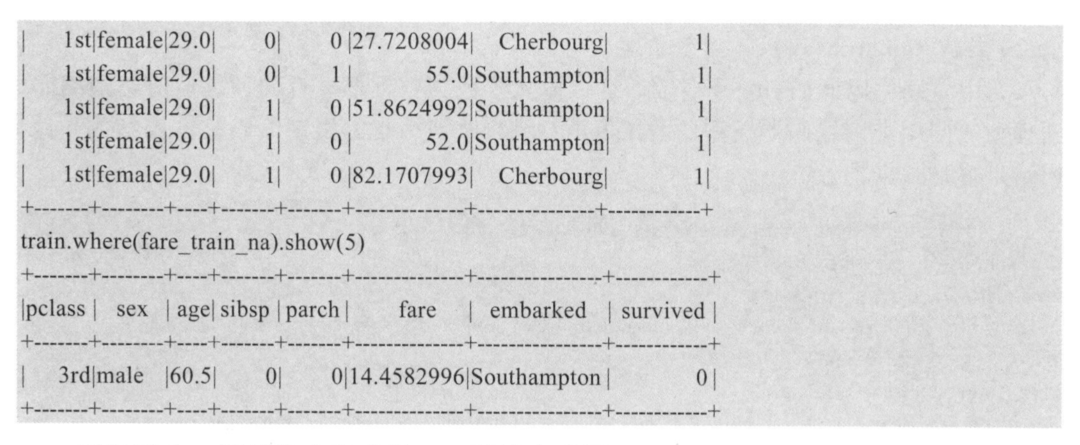

```
|    1st|female|29.0|      0|      0 |27.7208004|    Cherbourg|             1|
|    1st|female|29.0|      0|      1 |        55.0|Southampton|             1|
|    1st|female|29.0|      1|      0 |51.8624992|Southampton|             1|
|    1st|female|29.0|      1|      0 |        52.0|Southampton|             1|
|    1st|female|29.0|      1|      0 |82.1707993|    Cherbourg|             1|
+-------+--------+----+-------+-------+-------------+---------------+------------+
train.where(fare_train_na).show(5)
+-------+--------+----+-------+-------+-------------+---------------+------------+
|pclass |   sex   | age| sibsp | parch |      fare     |  embarked   | survived |
+-------+--------+----+-------+-------+-------------+---------------+------------+
|    3rd|male    |60.5|      0|      0|14.4582996|Southampton |             0|
+-------+--------+----+-------+-------+-------------+---------------+------------+
```

可以看出，训练集中的变量 age 都填充成了 29.0。

2. 填充测试集

做数据填充时，无论是用统计量还是机器学习模型，都必须用训练集中的信息。在做模型测试前对测试集做数据填充时，也需要用训练集的统计量或基于训练集训练的机器学习模型。

调用填充器的 transform()函数填充测试集。

```
test = impm.transform(test)
test.where(age_test_na).show(5)
+-------+--------+----+-------+-------+------------+---------------+------------+
|pclass |   sex   | age| sibsp |parch |      fare    |  embarked  |survived  |
+-------+--------+----+-------+-------+------------+---------------+------------+
|    1st|female|29.0|      0|      0|    31.6833|Southampton|             1|
|    1st|female|29.0|      0|      0|79.1999969|   Cherbourg|             1|
|    1st|female|29.0|      0|      0|110.883301|   Cherbourg|             1|
|    1st|female|29.0|      1|      0|89.1042023|   Cherbourg|             1|
|    1st|female|29.0|      1|      0|146.520798|   Cherbourg|             1|
+-------+--------+----+-------+-------+------------+---------------+------------+
```

可以看出，测试集中的变量 age 也都填充成了 29.0。

11.4 类别变量处理

可以看出，数据集中变量 pclass、sex 和 embarked 都是类别变量，即其值不是数值型。MLlib 中的建模函数多数只允许数值型变量，因此需要将类别型变量转换为数值型变量。例如，变量 pclass 中“1st”都转换为 1，“2nd”都转换为 2，“3rd”都转换为 0。

1. 类别编码

调用 pyspark.ml.feature 程序包中的 StringIndexer()构造函数创建 3 个类别编码器，即每个类别变量都需要一个，其中：

- 参数 inputCol 表示输入变量名；

• 参数 outputCol 表示输出变量名。

调用类别编码器的 fit()函数做训练，输入参数为数据框，返回训练后的类别编码器模型。调用类别编码器模型的 transform()函数做训练集的类别编码，输入参数和返回结果都是数据框。

```
si_pclass = StringIndexer(inputCol = 'pclass', outputCol = 'pclass_indexed')
sim_pclass = si_pclass.fit(train)
train = sim_pclass.transform(train)
si_sex = StringIndexer(inputCol = 'sex', outputCol = 'sex_indexed')
sim_sex = si_sex.fit(train)
train = sim_sex.transform(train)
si_embarked = StringIndexer(inputCol = 'embarked', outputCol = 'embarked_indexed')
sim_embarked = si_embarked.fit(train)
train = sim_embarked.transform(train)
train.show(5)
+------+---+----+-----+------+----+-----------+---------+-----------------+--------------+--------------------+
|pclass| sex| age|sibsp|parch| fare|embarked|survived|pclass_indexed|sex_indexed|embarked_indexed|
+---+----------+----+---+----+-----------------+--------------+--------+-----------+------------+--------------+
| 1st|   female|29.0|  0|  0|   27.7208004|   Cherbourg|       1|        1.0|        1.0|             1.0|
| 1st|   female|29.0|  0|  1|         55.0|Southampton|       1|        1.0|        1.0|             0.0|
| 1st|   female|29.0|  1|  0|   51.8624992|Southampton|       1|        1.0|        1.0|             0.0|
| 1st|   female|29.0|  1|  0|         52.0|Southampton|       1|        1.0|        1.0|             0.0|
| 1st|   female|29.0|  1|  0|   82.1707993|   Cherbourg|       1|        1.0|        1.0|             1.0|
+---+----------+----+---+----+-----------------+--------------+--------+-----------+------------+--------------+
```

调用类别编码器的 transform()函数做测试集的类别编码。

```
test = sim_pclass.transform(test)
test = sim_sex.transform(test)
test = sim_embarked.transform(test)
test.show(5)
+------+---+----+-----+------+----+------------+---------+-----------------+--------------+--------------------+
|pclass| sex| age|sibsp|parch| fare| embarked|survived|pclass_indexed|sex_indexed|embarked_indexed|
+---+---------+----+---+---+-----------------+----------------+--------+-----------+------------+--------------+
| 1st|  female|29.0|  0|  0|     31.6833|  Southampton|       1|        1.0|        1.0|             0.0|
| 1st|  female|29.0|  0|  0|  79.1999969|     Cherbourg|       1|        1.0|        1.0|             1.0|
| 1st|  female|29.0|  0|  0|  110.883301|     Cherbourg|       1|        1.0|        1.0|             1.0|
| 1st|  female|29.0|  1|  0|  89.1042023|     Cherbourg|       1|        1.0|        1.0|             1.0|
| 1st|  female|29.0|  1|  0|  146.520798|     Cherbourg|       1|        1.0|        1.0|             1.0|
+---+---------+----+---+---+-----------------+----------------+--------+-----------+------------+--------------+
```

可以看出，所有类别型变量都已经转换为数值型变量。对于多于 2 个独立类别的变量，这样的处理隐含了不同类别的序数关系。例如，使用类别编码器模型的属性 labels 得到每个序数对应的类别。

```
print(sim_embarked.labels)
## [u'Southampton', u'Cherbourg', u'Queenstown', u'Unknown']
print(sim_pclass.labels)
## [u'3rd', u'1st', u'2nd']
```

可以看出，0 对应 Southampton，1 对应 Cherbourg，2 对应 Queenstown，3 对应

Unknown。如果不加以处理，则隐含了序列关系，即 Southampton 小于 Cherbourg 小于 Queenstown，这显然不是我们想要的。

2．独热编码

独热编码（one-hot encoding）为每个独立值创建一个哑变量（dummy variable）。例如，变量 embarked 经过类别编码后，有 4 种独立值，分别为 0、1、2 和 3。独热编码会将这一变量变换为 4 个变量，对于值为 0 的记录编码为[1,0,0]，对于值为 1 的记录编码为[0,1,0]，依此类推。默认情况下，最后一个类别对应的哑变量不包含在内，以避免线性依赖，即有 N 个独立值的变量经过独热编码后，仅有 $N-1$ 个哑变量。

调用 pyspark.ml.feature 程序包中的 OneHotEncoderEstimator()构造函数创建独热编码器，其中：

- 参数 inputCols 表示输入变量名列表；
- 参数 outputCols 表示输出变量名列表。

调用独热编码器的 fit()函数做训练，输入参数为数据框，返回训练后的独热编码器模型。调用独热编码器模型的 transform()函数做训练集的独热编码，输入参数和返回结果都是数据框。

```
ohe = OneHotEncoderEstimator(inputCols = ['pclass_indexed', 'embarked_indexed'], outputCols =
['pclass_encoded', 'embarked_encoded'])
ohem = ohe.fit(train)
train = ohem.transform(train)
train.show(5)
+------+---+----+-----+------+----+------------+---------+----------------+-------------+--------------------+----
--------------------+---------------------+
|pclass| sex| age|sibsp|parch| fare| embarked|survived|pclass_indexed|sex_indexed|embarked_ indexed|
 pclass_encoded|embarked_encoded|
+---+-------+----+-+-+------------+-------------+-----+-------+------+------+-------------+--------------+
| 1st|female|29.0| 0| 0|27.7208004|   Cherbourg|     1|     1.0|    1.0|    1.0| (2,[1],[1.0])| (3,[1],[1.0])|
| 1st|female|29.0| 0| 1|        55.0|Southampton|     1|     1.0|    1.0|    0.0| (2,[1],[1.0])| (3,[0],[1.0])|
| 1st|female|29.0| 1| 0|51.8624992|Southampton|     1|     1.0|    1.0|    0.0| (2,[1],[1.0])| (3,[0],[1.0])|
| 1st|female|29.0| 1| 0|        52.0|Southampton|     1|     1.0|    1.0|    0.0| (2,[1],[1.0])| (3,[0],[1.0])|
| 1st|female|29.0| 1| 0|82.1707993|   Cherbourg|     1|     1.0|    1.0|    1.0| (2,[1],[1.0])| (3,[1],[1.0])|
+---+-------+----+-+-+------------+-------------+-----+-------+------+------+-------------+--------------+
```

调用独热编码器的 transform()函数做测试集的独热编码。

```
test = ohem.transform(test)
test.show(5)
+------+---+-----+-----+-----+----+------------+---------+----------------+-------------+--------------------+----
------------------+---------------------+
|pclass| sex| age|sibsp|parch| fare| embarked|survived|pclass_indexed|sex_indexed|embarked_indexed|
pclass_encoded|embarked_encoded|
+---+-------+----+-+-+------------+---------------+---+---------+-------+---+-------------+-------------+
| 1st|female|29.0| 0| 0|     31.6833|Southampton|  1|     1.0|     1.0| 0.0| (2,[1],[1.0])| (3,[0],[1.0])|
| 1st|female|29.0| 0| 0|79.1999969|   Cherbourg|  1|     1.0|     1.0| 1.0| (2,[1],[1.0])| (3,[1],[1.0])|
| 1st|female|29.0| 0| 0|110.883301|   Cherbourg|  1|     1.0|     1.0| 1.0| (2,[1],[1.0])| (3,[1],[1.0])|
| 1st|female|29.0| 1| 0|89.1042023|   Cherbourg|  1|     1.0|     1.0| 1.0| (2,[1],[1.0])| (3,[1],[1.0])|
```

```
| 1st|female|29.0| 1| 0|146.520798|   Cherbourg|   1|       1.0|      1.0| 1.0| (2,[1],[1.0])| (3,[1],[1.0])|
+---+-------+----+-+-+-------------+---------------+---+---------+-------+---+-------------+--------------+
```

11.5 特 征 选 择

1. 向量化

在建立机器学习模型前，需要将所有预测变量封装成向量。

调用 pyspark.ml.feature 程序包中的 VectorAssembler()构造函数创建向量封装器，其中：

- 参数 inputCols 表示输入预测变量名列表；
- 参数 outputCol 指定输出向量名。

调用向量封装器的函数 transform()做训练集的向量封装，将预测变量封装为名为 features 的向量，输入参数和返回结果都是数据框。

```
va = VectorAssembler(inputCols=['pclass_encoded','sex_indexed', 'age','sibsp','parch','fare','embarked_encoded'],
outputCol='features')
train = va.transform(train)
train.show(5)
+------+---+----+-----+-----+-----+-----------+---------+-----------------+-------------+---------------------+----
-----------------+---------------------+---------+
|pclass| sex| age|sibsp|parch| fare| embarked|survived|pclass_indexed|sex_indexed|embarked_indexed|
pclass_encoded|embarked_encoded| features|
+---+------+----+-+-+-------------+-----------------+-+-+--+---+-----------+-------------+------------------+
| 1st|female|29.0| 0| 0|27.7208004  | Cherbourg      |1|1.0|1.0|1.0|(2,[1],[1.0])|(3,[1],[1.0])|(10,[1,2,3,6,8],[... |
| 1st|female|29.0| 0| 1|55.0        |Southampton     |1|1.0|1.0|0.0|(2,[1],[1.0])|(3,[0],[1.0])|[0.0,1.0,1.0,29.0...|
| 1st|female|29.0| 1| 0|51.8624992  |Southampton     |1|1.0|1.0|0.0|2,[1],[1.0]) |(3,[0],[1.0])|[0.0,1.0,1.0,29.0...|
| 1st|female|29.0| 1| 0| 52.0       |Southampton     |1|1.0|1.0|0.0|(2,[1],[1.0])|(3,[0],[1.0])|[0.0,1.0,1.0,29.0...|
| 1st|female|29.0| 1| 0|82.1707993  | Cherbourg      |1|1.0|1.0|1.0|(2,[1],[1.0])|(3,[1],[1.0])|[0.0,1.0,1.0,29.0...|
+---+------+----+-+-+-------------+-----------------+-+-+--+---+-----------+-------------+------------------+
```

调用向量封装器的 transform()函数做测试集的向量封装。

```
test = va.transform(test)
test.show(5)
+-----+---+----+----+-----+----+----------+---------+--------------+------------+------------------+-------------
---------+--------------------+------------+
|pclass| sex| age|sibsp|parch| fare| embarked|survived|pclass_indexed|sex_indexed|embarked_indexed|pclass_
encoded|embarked_encoded|    features|
+--+-------+----+-+--+------------+---------------+-+-+---+---+------------+------------+--------------------+
|1st|female|29.0|0 |0 |31.6833      |Southampton|1|1.0|1.0|0.0|(2,[1],[1.0])|(3,[0],[1.0])|(10,[1,2,3,6,7],[... |
|1st|female|29.0|0 |0 |79.1999969 |Cherbourg    |1|1.0|1.0|1.0|(2,[1],[1.0])|(3,[1],[1.0])|(10,[1,2,3,6,8],[... |
|1st|female|29.0|0 |0 |110.883301 |Cherbourg    |1|1.0|1.0|1.0|(2,[1],[1.0])|(3,[1],[1.0])|(10,[1,2,3,6,8],[... |
|1st|female|29.0|1 |0 |89.1042023 |Cherbourg    |1|1.0|1.0|1.0|(2,[1],[1.0])|(3,[1],[1.0])|[0.0,1.0,1.0,29.0...|
|1st|female|29.0|1 |0 |146.520798 |Cherbourg    |1|1.0|1.0|1.0|(2,[1],[1.0])|(3,[1],[1.0])|[0.0,1.0,1.0,29.0...|
+--+-------+----+-+--+------------+---------------+-+-+---+---+------------+------------+--------------------+
```

虽然目标变量 survived 已经是数值型变量，但缺乏机器学习模型所需要的元数据（如类别数量、独立值等），仍然需要调用类别编码器做变换。

```
si_survived = StringIndexer(inputCol = 'survived', outputCol = 'survived_indexed')
sim_survived = si_survived.fit(train)
train = sim_survived.transform(train)
train.show(5)
+------+---+----+-----+-----+-----+-----------+---------+----------------+-------------+---------------------+-----
------------------+---------------------+--------+--------------------+
|pclass| sex| age|sibsp|parch| fare|embarked|survived|pclass_indexed|sex_indexed|embarked_indexed|
pclass_encoded|embarked_encoded|features|survived_indexed|
+--+-------+---+-+-+-----------+----------------+-+--+--+---+-----------+------------+-------------------+--+
|1st|female|29.0|0 |0 |27.7208004|Cherbourg    |1 |1.0|1.0|1.0|(2,[1],[1.0])|(3,[1],[1.0])|(10,[1,2,3,6,8],[...|1.0|
|1st|female|29.0|0 |1 |55.0      |Southampton  |1 |1.0|1.0|0.0|(2,[1],[1.0])|(3,[0],[1.0])|[0.0,1.0,1.0,29.0...|1.0|
|1st|female|29.0|1 |0 |51.8624992|Southampton  |1 |1.0|1.0|0.0|(2,[1],[1.0])|(3,[0],[1.0])|[0.0,1.0,1.0,29.0...|1.0|
|1st|female|29.0|1 |0 |52.0      |Southampton  |1 |1.0|1.0|0.0|(2,[1],[1.0])|(3,[0],[1.0])|[0.0,1.0,1.0,29.0...|1.0|
|1st|female|29.0|1 |0 |82.1707993|Cherbourg    |1 |1.0|1.0|1.0|(2,[1],[1.0])|(3,[1],[1.0])|[0.0,1.0,1.0,29.0...|1.0|
+--+-------+---+-+-+-----------+----------------+-+--+--+---+-----------+------------+-------------------+--+
test = sim_survived.transform(test)
test.show(5)
+-----+----+---+-----+-----+----+----------+---------+---------------+------------+-------------------+-----------
---------+---------------------+----------+-------------------+--------------------+
|pclass| sex| age|sibsp|parch| fare| embarked|survived|pclass_indexed|sex_indexed|embarked_indexed|pclass_
encoded|embarked_encoded| features|survived_indexed|
+--+-------+----+-+-+------------+-------------+-+-+--+---+---+------------+-----------+--------------------+--+
|1st|female|29.0|0|0 |31.6833    |Southampton|1 |1.0|1.0|0.0|(2,[1],[1.0])|(3,[0],[1.0])|(10,[1,2,3,6,7],[... |1.0|
|1st|female|29.0|0|0 |79.1999969|Cherbourg    |1 |1.0|1.0|1.0|(2,[1],[1.0])|(3,[1],[1.0])|(10,[1,2,3,6,8],[... |1.0|
|1st|female|29.0|0|0 |110.883301|Cherbourg    |1 |1.0|1.0|1.0|(2,[1],[1.0])|(3,[1],[1.0])|(10,[1,2,3,6,8],[... |1.0|
|1st|female|29.0|1|0 |89.1042023|Cherbourg    |1 |1.0|1.0|1.0|(2,[1],[1.0])|(3,[1],[1.0])|[0.0,1.0,1.0,29.0...|1.0|
|1st|female|29.0|1|0 |146.520798|Cherbourg    |1 |1.0|1.0|1.0|(2,[1],[1.0])|(3,[1],[1.0])|[0.0,1.0,1.0,29.0...|1.0|
+--+-------+----+-+-+------------+-------------+-+-+--+---+---+------------+-----------+--------------------+--+
```

只保留预测变量和目标变量，并将目标变量重命名为 label。

```
train = train.select(train.features, train.survived_indexed.alias ('label'))
train.show(5)
+--------------------+------+
|      features      |label |
+--------------------+------+
|(10,[1,2,3,6,8],[... |  1.0|
|[0.0,1.0,1.0,29.0...|  1.0|
|[0.0,1.0,1.0,29.0...|  1.0|
|[0.0,1.0,1.0,29.0...|  1.0|
|[0.0,1.0,1.0,29.0...|  1.0|
+--------------------+------+
test = test.select(test.features, test.survived_indexed.alias ('label'))
test.show(5)
+---------------------+-----+
|       features      |label |
+---------------------+-----+
|(10,[1,2,3,6,7],[... |  1.0|
|(10,[1,2,3,6,8],[... |  1.0|
```

```
|(10,[1,2,3,6,8],[... |  1.0|
|[0.0,1.0,1.0,29.0...|  1.0|
|[0.0,1.0,1.0,29.0...|  1.0|
+--------------------+-----+
```

在做完独热编码后，列索引对应的变量意义如下：

- 列索引 0－1：变量 pclass 是否为 3rd 和 1st，1 为是，0 为否；
- 列索引 2：变量 sex，0 为 male，1 为 female；
- 列索引 3－6：变量 age、sibsp、parch 和 fare。
- 列索引 7－9：变量 embarked 是否为 Southampton、Cherbourg 和 Queenstown，1 为是，0 为否。

定义列索引对应的变量名。

```
features = ['pclass_3','pclass_1','sex','age','sibsp','parch', 'fare','embarked_S','embarked_C','embarked_Q']
```

2. 卡方特征选择

特征选择是指在数据集中选取若干代表性强或预测能力强的变量子集，使得机器学习模型训练更高效且性能不下降。

调用 pyspark.ml.feature 程序包中的 ChiSqSelector()构造函数创建卡方特征选择器，用于选择重要性最高的 5 个特征。其中：

- 第 1 个参数 numTopFeatures 表示选择重要性最高的特征变量数；
- 第 2 个参数 featuresCol 表示输入向量名，默认为 features；
- 第 3 个参数 outputCol 表示输出向量名；
- 第 4 个参数 labelCol 表示目标变量名，默认为 label。

调用卡方特征选择器的 fit()函数做训练，输入参数为数据框，返回训练后的卡方特征选择器模型。调用卡方特征选择器模型的 transform()函数做训练集的特征选择，输入参数和返回结果都是数据框。

```
css = ChiSqSelector(numTopFeatures=5, outputCol="selectedFeatures")
cssm = css.fit(train)
train = cssm.transform(train)
train.show(5)
+--------------------+------+--------------------+
|          eatures   |label |   selectedFeatures|
+--------------------+------+--------------------+
|(10,[1,2,3,6,8],[... |  1.0|(5,[1,2,3,4],[1.0... |
|[0.0,1.0,1.0,29.0...|  1.0|[0.0,1.0,1.0,55.0...|
|[0.0,1.0,1.0,29.0...|  1.0|[0.0,1.0,1.0,51.8...|
|[0.0,1.0,1.0,29.0...|  1.0|[0.0,1.0,1.0,52.0...|
|[0.0,1.0,1.0,29.0...|  1.0|[0.0,1.0,1.0,82.1...|
+--------------------+------+--------------------+
```

可以看出，输出向量仅保留了 5 个变量。

使用特征选择器模型的属性 selectedFeatures 得到选择特征的序号，并进一步得到变量名称。

```
print(cssm.selectedFeatures)
##[2, 0, 1, 6, 8]
[features[i] for i in cssm.selectedFeatures]
##['sex', 'pclass_3', 'pclass_1', 'fare', 'embarked_C']
```

可以看出，变量 sex、pclass、fare 和 embarked 的重要性较高。

11.6 建模与调优

11.6.1 创建机器学习模型

机器学习建模有多种任务，包括分类预测、回归预测和聚类分析等。每种任务又都有多种模型，如对于分类任务，有决策树、随机森林、支持向量机和人工神经网络等。

这里调用 pyspark.ml.classification 程序包中的 DecisionTreeClassifier( )构造函数创建决策树分类器，之后章节会具体介绍。其中：

- 参数 featuresCol 表示预测变量向量名，默认为 features；
- 参数 labelCol 表示目标变量名，默认为 label。

```
dtc = DecisionTreeClassifier( )
```

11.6.2 超参数调优

对于每种模型，都有一系列超参数，即需要在训练模型前人工指定模型参数。

调用决策树分类器的 explainParams()函数得到超参数列表和每个超参数的默认值。

```
print(dtc.explainParams( ))
cacheNodeIds: If false, the algorithm will pass trees to executors to match instances with nodes. If true, the algorithm will cache node IDs for each instance. Caching can speed up training of deeper trees. Users can set how often should the cache be checkpointed or disable it by setting checkpointInterval. (default: False)
checkpointInterval: set checkpoint interval (>= 1) or disable checkpoint (-1). E.g. 10 means that the cache will get checkpointed every 10 iterations. Note: this setting will be ignored if the checkpoint directory is not set in the SparkContext. (default: 10)
featuresCol: features column name. (default: features)
impurity: Criterion used for information gain calculation (case-insensitive). Supported options: entropy, gini (default: gini)
labelCol: label column name. (default: label)
maxBins: Max number of bins for discretizing continuous features.  Must be >=2 and >= number of categories for any categorical feature. (default: 32)
maxDepth: Maximum depth of the tree. (>= 0) E.g., depth 0 means 1 leaf node; depth 1 means 1 internal node + 2 leaf nodes. (default: 5)
maxMemoryInMB: Maximum memory in MB allocated to histogram aggregation. If too small, then 1 node will be split per iteration, and its aggregates may exceed this size. (default: 256)
minInfoGain: Minimum information gain for a split to be considered at a tree node. (default: 0.0)
minInstancesPerNode: Minimum number of instances each child must have after split. If a split causes the left or right child to have fewer than minInstancesPerNode, the split will be discarded as invalid. Should be >= 1. (default: 1)
```

```
predictionCol: prediction column name. (default: prediction)
probabilityCol: Column name for predicted class conditional probabilities. Note: Not all models output well-calibrated probability estimates! These probabilities should be treated as confidences, not precise probabilities. (default: probability)
rawPredictionCol: raw prediction (a.k.a. confidence) column name. (default: rawPrediction)
seed: random seed. (default: 6174923023070228847)
```

对于具体问题，无法事先明确知道哪一组超参数会取得最佳效果，因此需要做超参数调优得到最佳的超参数组合。

超参数调优需要指定以下几个选项：超参数的搜索范围；调优算法；评估方法，即重采样策略和评估指标。

1. 定义超参数调优算法

调用 pyspark.ml.tuning 程序包中的 ParamGridBuilder()构造函数创建网格搜索超参数调优器。

```
pgb = ParamGridBuilder()
```

2. 定义超参数的搜索范围

调用网格搜索超参数调优器的 addGrid()函数添加需要调优的超参数及尝试的数值。其中：

- 第 1 个参数 param 表示需要调优的超参数；
- 第 2 个参数 values 表示尝试的数值列表。

这里调优的超参数范围为：

- 最大深度（maxDepth）：3、4 和 5；
- 最小信息增益（minInfoGain）：0.001 和 0.005。

调用网格搜索超参数调优器的 build()函数创建超参数字典。

```
grid = pgb.addGrid(dtc.maxDepth, [3,4,5]).addGrid(dtc.minInfoGain, [0.001,0.005]).build()
```

3. 定义重采样策略

重采样指的是将训练集进一步划分成训练集和验证集的方法。常用的策略有：

- 留出法（hold-out）：将数据集划分成互斥的训练集和验证集；
- 交叉验证法（cross validation）：将数据集划分成 k 个大小相似的互斥子集，每次用 $k-1$ 个子集的并集作为训练集，余下的那个子集作为验证集，从而进行 k 次训练和验证，最终返回的是这 k 个验证集结果的均值，如图 11-3 所示（$k=5$）；
- 余一交叉验证法（leave-one-out）：将交叉验证法中的折数 k 设为样本数 m。

调用 pyspark.ml.tuning 程序包中的 CrossValidator()构造函数创建交叉验证器。

- 第 1 个参数 estimator 表示需要做超参数调优的模型；
- 第 2 个参数 estimatorParamMaps 表示超参数字典；
- 第 3 个参数 evaluator 表示模型性能评估器，之后章节会具体介绍；
- 第 4 个参数 numFolds 表示交叉验证中的折数。

```
cv = CrossValidator(estimator = dtc, evaluator = BinaryClassification Evaluator(),
  estimatorParamMaps = grid, numFolds = 3)
```

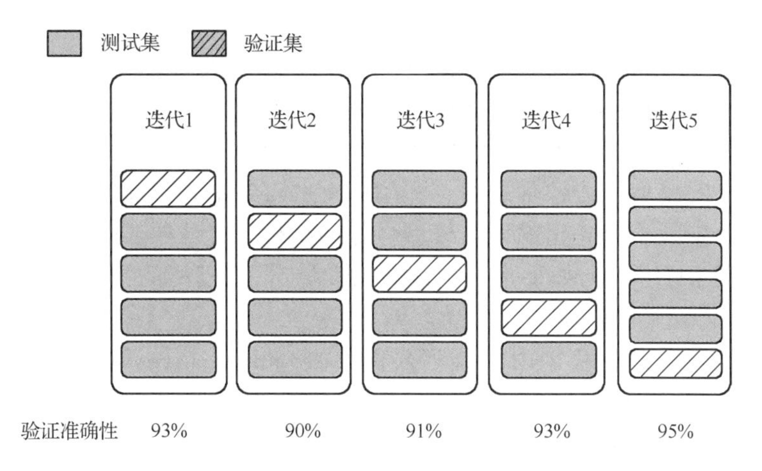

图 11-3　交叉验证法（ $k=5$ ）

4. 执行超参数调优

调用交叉验证器的 fit()函数做超参数调优，输入参数为数据框，返回训练后的交叉验证器模型。

使用交叉验证器模型的 bestModel 属性得到最优的机器学习模型，并使用其属性 _java_obj 中的相应函数 get<超参数名称>()得到最佳超参数组合是最大深度（maxDepth）为 3 且最小掺杂度减少比例（minInfoGain）为 0.001。

```
cvm = cv.fit(train)
print "maxDepth = ", cvm.bestModel._java_obj.parent( ).getMaxDepth( )
## maxDepth =    3
print "minInfoGain = ", cvm.bestModel._java_obj.parent( ).get MinInfoGain( )
## minInfoGain =    0.001
```

11.7　测试与评估

在训练集上做完超参数调优和模型训练后，需要在测试集上对模型做一次性公正的评估。

调用决策树分类器模型的 transform()函数在测试集上预测类别标签和概率，返回结果中：

- 变量 rawPrediction 表示模型预测的原始结果，对于决策树分类模型，为叶子节点中两类的数量；
- 变量 probability 表示两类的预测概率；
- 变量 prediction 表示预测的类别标签（0.0 或 1.0）。

```
prediction = cvm.bestModel.transform(test)
prediction.show(5)
+-------------------+----+--------------+------------------------+-------------+
|        features   |label|rawPrediction|        probability     | prediction |
+-------------------+----+--------------+------------------------+-------------+
|(10,[1,2,3,6,7],[... | 1.0 | [10.0,154.0]|[0.06097560975609...|          1.0|
|(10,[1,2,3,6,8],[... | 1.0 | [10.0,154.0]|[0.06097560975609...|          1.0|
```

```
|(10,[1,2,3,6,8],[... | 1.0 | [10.0,154.0]|[0.06097560975609...|            1.0|
|[0.0,1.0,1.0,29.0...| 1.0 | [10.0,154.0]|[0.06097560975609...|            1.0|
|[0.0,1.0,1.0,29.0...| 1.0 | [10.0,154.0]|[0.06097560975609...|            1.0|
+--------------------+----+--------------+------------------------+-------------+
```

在做更复杂的模型性能评估之前，先了解一些直观的性能指标。

精度并不一定是一个公正的衡量模型性能的指标。试想对于 100 个样本，其中 99 个阴性样本，1 个阳性样本，如果全部预测为阴性样本，则模型的精度也可以达到 99%，然而这样的模型并不一定是我们想要的。

对于二分类问题，常用 ROC 曲线下面积（area under curve，AUC）作为模型性能指标，在介绍该指标之前，首先需要定义如下 4 种预测结果：

- 真阳性（true positive，TP）：预测为阳性，实际为阳性；
- 假阳性（false positive，FP）：预测为阳性，实际为阴性；
- 真阴性（true negative，TN）：预测为阴性，实际为阴性；
- 假阴性（false negative，FN）：预测为阴性，实际为阳性。

也可以通过表 11-2 呈现，并定义真阳性率（tpr）和假阳性率（fpr）。

表 11-2　二分类问题预测结果表

	预测为阳性	预测为阴性	
实际为阳性	真阳性（tp）	假阴性（fn）	真阳性率（tpr），tp/(tp+fn)
实际为阴性	假阳性（fp）	真阴性（tn）	假阳性率（fpr），fp/(fp+tn)

另外常用的还有查准率（precision）、查全率（recall）和综合了两者的 F1 度量。

$$\text{precision} = \frac{\text{tp}}{\text{tp} + \text{tp}}$$

$$\text{recall} = \frac{\text{tp}}{\text{tp} + \text{fn}}$$

$$\text{F1} = \frac{2 \times \text{precision} \times \text{recall}}{\text{precision} + \text{recall}}$$

受试者工作特征曲线（receiver operating characteristic curve，ROC 曲线），即假阳性率（fpr）为横轴、真阳性率（tpr）为纵轴所画出的曲线。分类阈值分别取 0 到 1 中的各个数，计算不同阈值对应的 fpr 和 tpr，描点得到 ROC 曲线。

ROC 曲线下面积（auc）会出现如下 3 种情况：

- $0.5 < \text{auc} \leqslant 1$ 时，模型优于随机猜测，设定合适的阈值则具有预测价值；
- $\text{auc} = 0.5$ 时，模型与随机猜测一致，无预测价值；
- $0 \leqslant \text{auc} < 0.5$ 时，模型差于随机猜测，但如果总是反预测而行，反而能产生预测价值。

计算真阳性（tp）、假阳性（fp）、真阴性（tn）和假阴性（fn）样本数，并进一步计算模型精度、查准率和查全率。

```
tp = float(prediction.filter("prediction = 1.0 AND label = 1").count( ))
fp = float(prediction.filter("prediction = 1.0 AND label = 0").count( ))
```

```
tn = float(prediction.filter("prediction = 0.0 AND label = 0").count( ))
fn = float(prediction.filter("prediction = 0.0 AND label = 1").count( ))
accuracy = (tp + tn) / (tp + fp + tn + fn)
print(accuracy)
## 0.827067669173
precision = tp / (tp + fp)
print(precision)
## 0.765100671141
recall = tp / (tp + fn)
print(recall)
## 0.77027027027
```

调用 pyspark.ml.evaluation 程序包中的 BinaryClassificationEvaluator()构造函数创建二分类评估器。其中：

- 第 1 个参数 labelCol 表示真实类别标签的变量名称，默认为 label；
- 第 2 个参数 rawPredictionCol 表示预测概率的变量名称，默认为 rawPrediction；
- 第 3 个参数 metricName 表示模型性能指标名称，常用值为 areaUnderROC，表示 ROC 曲线下面积。

调用二分类评估器的 evaluate()函数计算模型性能指标。

```
bce = BinaryClassificationEvaluator(metricName ='areaUnderROC')
auc = bce.evaluate(prediction)
print(auc)
## 0.667290298266
```

小　结

本章主要介绍了在 Spark 中做机器学习，包括数据探索、数据划分、数据填充、特征选择、建模与调优以及测试与评估。数据探索为第一步，了解数据集中各变量的分布和缺失值情况等。数据划分将数据集分为训练集和测试集，防止模型的过拟合。数据填充主要用于填充数据集中的缺失值，避免一些模型由于缺失值而无法拟合。特征选择指的是选择数据集中有预测能力的变量，使得训练模型更加高效。建模与调优为机器学习的核心，即训练机器学习模型，并调整超参数得到最佳的性能。测试与评估通过一系列指标对模型性能做出公正的评价。

本章使用 Spark 中的 MLlib 实现机器学习的各个环节，其优点为提供了一系列高层的数据分析函数，如超参数调优、数据填充等便捷函数。

习　题

1. 机器学习的主要流程分为哪些步骤？
2. 训练集、验证集和测试集的用法和区别主要有哪些？
3. 基于泰坦尼克号乘客数据集，尝试将变量 age 的缺失值都填充为 30。
4. 做类别编码和独热编码的作用是什么？试用例子证明不做类别变量处理的危害。
5. 基于泰坦尼克号乘客数据集，尝试调优超参数 minInstancesPerNode，并以 F1 作为超参数调优的模型性能指标。

第12章 Spark 有监督学习模型

本章主要介绍使用 Spark 中的 pyspark.ml（ML）程序包实现常用的有监督学习模型。

有监督学习定义为，样本标签信息已知且提供给训练模型。如果标签信息是数值类型，则为回归问题；如果是类别类型，则为分类问题。这是两种最常见的有监督学习问题，也是本章介绍的重点。

载入本章需要用到的程序包。

```
from pyspark.ml.regression import LinearRegression, RandomForestRegressor, \
  DecisionTreeRegressor
from pyspark.ml.classification import LogisticRegression, \
  DecisionTreeClassifier, RandomForestClassifier, MultilayerPerceptron Classifier
from pyspark.ml.evaluation import RegressionEvaluator, \
  BinaryClassificationEvaluator, MulticlassClassificationEvaluator
from pyspark.ml.feature import VectorAssembler, StringIndexer
from pyspark.ml.tuning import ParamGridBuilder, CrossValidator
from pyspark.ml.recommendation import ALS
```

本章所有例子基于 3 个数据集。第 1 个数据集是 1970 年波士顿各地区的房价和其他统计信息，包含 506 个样本，每个样本包含 14 个属性，如表 12-1 所示

表 12-1　波士顿房价数据集属性定义

属　性	定　义
crim	犯罪率
zn	住宅用地比例
indus	非零售商业用地比例
chas	是否临河，1 为是，0 为否
nox	氮氧化物浓度
rm	每栋住宅平均房间数
age	1940 年以前建筑比例
dis	与市中心的距离
rad	高速公路的可接近性
tax	房地产税率

续表

属　性	定　义
ptratio	师生比例
b	有色人种比例
lstat	低端人口比例
medv	房价中位数

该数据集的路径为/root/data/boston/boston.csv。

该数据集用于回归问题，因变量为房价中位数（变量 medv）。该数据集包含多达 13 个自变量，也可以用于检验模型区分变量重要性的能力。

调用 spark 的 read()函数载入波士顿房价数据集 boston。

```
boston = spark.read.csv('file:///root/data/boston/boston.csv',
  header='true', inferSchema='true')
boston.show(5)
+--------+----+------+----+-----+-------+----+-------+---+----+---------+------+-----+-----+
|    crim|  zn|indus|chas|  nox|     rm| age|    dis |rad| tax |  ptratio |     b |lstat |medv|
+--------+----+-----+-----+-----+-------+----+-------+---+----+---------+------+-----+-----+
|0.00632|18.0| 2.31|     0|0.538|6.575 |65.2|   4.09|   1|296|     15.3| 396.9 | 4.98|24.0|
|0.02731| 0.0 | 7.07|     0|0.469|6.421 |78.9|4.9671|   2|242|     17.8| 396.9 | 9.14|21.6|
|0.02729| 0.0 | 7.07|     0|0.469|7.185 |61.1|4.9671|   2|242|     17.8|392.83| 4.03|34.7|
|0.03237| 0.0 | 2.18|     0|0.458|6.998 |45.8|6.0622|   3|222|     18.7|394.63| 2.94|33.4|
|0.06905| 0.0 | 2.18|     0|0.458|7.147 |54.2|6.0622|   3|222|     18.7| 396.9 | 5.33|36.2|
+--------+----+-----+-----+-----+-------+----+-------+---+----+---------+-------+-----+----+
```

做预测变量的向量封装。

```
va = VectorAssembler(
  inputCols=['crim','zn','indus','chas','nox','rm','age','dis','rad','tax','ptratio','b','lstat'], outputCol = 'features')
boston = va.transform(boston)
```

只保留预测变量和目标变量，并将目标变量重命名为 label。

```
boston = boston.select(boston.features, boston.medv.alias('label'))
boston.show(5)
+--------------------+-----+
|          features   |label |
+--------------------+-----+
|[0.00632,18.0,2.3...| 24.0|
|[0.02731,0.0,7.07...| 21.6|
|[0.02729,0.0,7.07...| 34.7|
|[0.03237,0.0,2.18...| 33.4|
|[0.06905,0.0,2.18...| 36.2|
+--------------------+-----+
```

第 2 个数据集是双螺旋结构数据，为人工合成，包含 200 个样本，每个样本包含 3 个属性，如表 12-2 所示。

表 12-2 双螺旋结构数据集属性定义

属性	定义
class	类别
x	x 轴坐标
y	y 轴坐标

该数据集的路径为/root/data/spirals/spirals.csv。

该数据集用于二分类问题，因变量为类别（变量 class）。该数据集两类样本的决策边界高度非线性，也可以用于检验分类模型学习非线性决策边界的能力。

调用 spark 的 read()函数载入双螺旋结构数据 spirals。

```
spirals = spark.read.csv('file:///root/data/spirals/spirals.csv',
  header='true', inferSchema='true')
spirals.show(5)
+------+-------------+--------------+
|class |       x     |       y      |
+------+-------------+--------------+
|     1|0.333333333|           0.0|
|     1|0.343933096|0.043448854|
|     1|0.348689938|0.089528359|
|     1|0.347116555|0.137433166|
|     1|0.338838583|0.186278087|
+------+-------------+--------------+
```

做预测变量的向量封装。

```
va = VectorAssembler(inputCols=['x','y'], outputCol = 'features')
spirals = va.transform(spirals)
```

做目标变量的类别编码。

```
si_class = StringIndexer(inputCol = 'class', outputCol = 'label')
sim_class = si_class.fit(spirals)
spirals = sim_class.transform(spirals)
```

只保留预测变量和目标变量。

```
spirals = spirals.select(spirals.features, spirals.label)
spirals.show(5)
+---------------------+------+
|         features    |label |
+---------------------+------+
|[0.333333333,0.0]    |   1.0|
|[0.343933096,0.04...|   1.0|
|[0.348689938,0.08...|   1.0|
|[0.347116555,0.13...|   1.0|
|[0.338838583,0.18...|   1.0|
+---------------------+------+
```

双螺旋结构数据如图 12-1 所示。

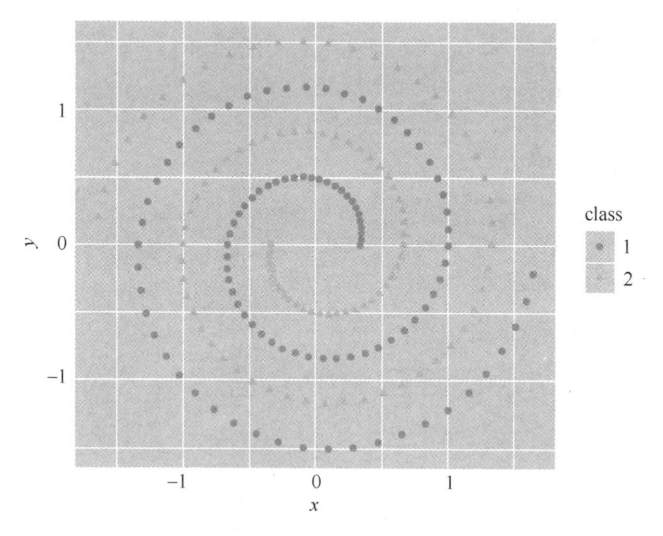

图 12-1 双螺旋结构（彩色图片见插页）

第 3 个数据集是鸢尾花卉数据集，由生物学家 Fisher 于 1936 年收集整理，包含 150 个样本，每个样本包含 5 个属性，如表 12-3 所示。

表 12-3 鸢尾花卉数据集属性定义

属　　性	定　　义
sepal_length	花萼长度
sepal_width	花萼宽度
petal_length	花瓣长度
petal_width	花瓣宽度
species	物种

该数据集的路径为/root/data/iris/iris.csv。

该数据集用于多分类问题，因变量为物种（变量 Species），其中包含了 3 个类别，0 为 Setosa，1 为 Versicolour，2 为 Virginica，每类 50 个样本。

调用 spark 的 read()函数载入鸢尾花卉数据集 iris。

```
iris = spark.read.csv('file:///root/data/iris/iris.csv',
  header='true', inferSchema='true')
iris.show(5)
+------------+-----------+------------+-----------+-------+
|sepal_length|sepal_width|petal_length|petal_width|species|
+------------+-----------+------------+-----------+-------+
|         5.1|        3.5|         1.4|        0.2|      0|
|         4.9|        3.0|         1.4|        0.2|      0|
|         4.7|        3.2|         1.3|        0.2|      0|
|         4.6|        3.1|         1.5|        0.2|      0|
|         5.0|        3.6|         1.4|        0.2|      0|
+------------+-----------+------------+-----------+-------+
```

做预测变量的向量封装。

```
va = VectorAssembler(inputCols=['sepal_length','sepal_width', 'petal_length','petal_width'], outputCol =
'features')
iris = va.transform(iris)
```

做目标变量的类别编码。

```
si_species = StringIndexer(inputCol = 'species', outputCol = 'label')
sim_species = si_species.fit(iris)
iris = sim_species.transform(iris)
```

只保留预测变量和目标变量。

```
iris = iris.select(iris.features, iris.label)
iris.show(5)
+------------------+-----+
|     features     |label |
+------------------+-----+
|[5.1,3.5,1.4,0.2] |  0.0|
|[4.9,3.0,1.4,0.2] |  0.0|
|[4.7,3.2,1.3,0.2] |  0.0|
|[4.6,3.1,1.5,0.2] |  0.0|
|[5.0,3.6,1.4,0.2] |  0.0|
+------------------+-----+
```

之后的内容将依次介绍一些常用的有监督学习模型，重点会覆盖：

- 模型在回归、两分类和多分类问题中的应用；
- 模型在 pyspark.ml 程序包中的实现方法；
- 模型各超参数的含义和影响。

12.1 线性回归模型

给定数据集 $D=\{(\boldsymbol{x}_1,y_1),(\boldsymbol{x}_2,y_2),\cdots,(\boldsymbol{x}_m,y_m)\}$，其中 $\boldsymbol{x}_i=(x_{i1};x_{i2};\cdots;x_{in}),y_i\in\mathbb{R}$。线性回归模型表达形式为：

$$f(\boldsymbol{x}_i)=\boldsymbol{w}^{\mathrm{T}}\boldsymbol{x}_i+b,\text{使得}f(\boldsymbol{x}_i)\approx y_i$$

线性回归一般仅用于回归问题，而不用于分类问题。

调用 pysprk.ml.regression 程序包中的 LinearRegression( )构造函数创建线性回归模型。得到超参数列表和每个超参数的默认值。

```
lr = LinearRegression( )
print(lr.explainParams( ))
aggregationDepth: suggested depth for treeAggregate (>= 2). (default: 2)
elasticNetParam: the ElasticNet mixing parameter, in range [0, 1]. For alpha = 0, the penalty is an L2
penalty. For alpha = 1, it is an L1 penalty. (default: 0.0)
epsilon: The shape parameter to control the amount of robustness. Must be > 1.0. Only valid when loss is
huber (default: 1.35)
featuresCol: features column name. (default: features)
fitIntercept: whether to fit an intercept term. (default: True)
```

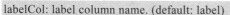

```
labelCol: label column name. (default: label)
loss: The loss function to be optimized. Supported options: squaredError, huber. (default: squaredError)
maxIter: max number of iterations (>= 0). (default: 100)
predictionCol: prediction column name. (default: prediction)
regParam: regularization parameter (>= 0). (default: 0.0)
solver: The solver algorithm for optimization. Supported options: auto, normal, l-bfgs. (default: auto)
standardization: whether to standardize the training features before fitting the model. (default: True)
tol: the convergence tolerance for iterative algorithms (>= 0). (default: 1e-06)
weightCol: weight column name. If this is not set or empty, we treat all instance weights as 1.0. (undefined)
```

可以看出，有多个可调优的超参数，最常用的包括：

- 参数 regParam：规则化参数；
- 参数 solver：优化算法，l-bfgs 为有限内存半牛顿优化法，normal 为解析法，auto 为自动选择。

1．超参数调优

创建网格搜索超参数调优器，添加需要调优的超参数及尝试的数值，这里调优的超参数为：

- 规则化参数（regParam）：0.1、0.2 和 0.5；
- 优化算法（solver）：l-bfgs 和 normal。

```
grid = ParamGridBuilder().addGrid(lr.regParam,[0.1,0.2,0.5]) \
   .addGrid(lr.solver,['l-bfgs','normal']).build()
```

做 3 折交叉验证，得到最佳超参数组合规则化参数（regParam）为 0.1 且优化算法（solver）为解析法（normal）。

```
cv = CrossValidator(estimator = lr, evaluator = RegressionEvaluator(),
   estimatorParamMaps = grid, numFolds = 3)
cvm = cv.fit(boston)
print "regParam = ", cvm.bestModel._java_obj.parent().getRegParam()
## regParam =    0.1
print "solver = ", cvm.bestModel._java_obj.parent().getSolver()
## solver =    normal
```

2．模型评估

对于回归问题，最常用 R^2（r2_score）作为模型性能指标。

定义因变量的均值为：

$$\bar{y} = \frac{1}{m}\sum_{i=1}^{m} y_i$$

定义因变量的平方和为

$$SS_{tot} = \frac{1}{m}\sum_{i=1}^{m}(y_i - \bar{y})^2$$

定义残差的平方和为：

$$\mathrm{SS}_{\mathrm{res}}=\frac{1}{m}\sum_{i=1}^{m}(y_i-f(\boldsymbol{x}_i))^2$$

则 R^2 定义为：

$$R^2=1-\frac{\mathrm{SS}_{\mathrm{res}}}{\mathrm{SS}_{\mathrm{tot}}}$$

可以理解为，模型能够解释的因变量变化情况的比例。另外，均方误差 $\mathrm{SS}_{\mathrm{res}}$（mean_squared_error）和均方根误差 $\sqrt{\mathrm{SS}_{\mathrm{res}}}$ 也比较常用。

调用线性回归模型的 transform()函数预测数值标签，这里为了突出模型，并没有分割训练集和测试集。

```
prediction = cvm.bestModel.transform(boston)
prediction.show(5)
+--------------------+-----+------------------+
|            features|label|        prediction|
+--------------------+-----+------------------+
|[0.00632,18.0,2.3...| 24.0| 30.11724819751664|
|[0.02731,0.0,7.07...| 21.6| 25.01573591016486|
|[0.02729,0.0,7.07...| 34.7|30.574180230668993|
|[0.03237,0.0,2.18...| 33.4|28.690012117562073|
|[0.06905,0.0,2.18...| 36.2|28.049058047467895|
+--------------------+-----+------------------+
```

调用 pyspark.ml.evaluation 程序包中的 RegressionEvaluator( )构造函数创建回归评估器。其中：

- 第 1 个参数 predictionCol 表示预测值的变量名称，默认为 prediction；
- 第 2 个参数 labelCol 表示真实值的变量名称，默认为 label；
- 第 3 个参数 metricName 表示模型性能指标名称，默认为 rmse，表示均方根误差。

调用二分类评估器的 evaluate()函数计算模型性能指标。其中：

- 第 1 个参数 dataset 表示预测结果；
- 第 2 个参数 params 表示覆盖原评估器的参数字典，当覆盖评估器的参数 metricName 为'r2'，表示 R^2；为'mse'，表示均方误差。

```
re = RegressionEvaluator( )
re.evaluate(prediction, {re.metricName:'r2'})
## 0.7403355590299989
re.evaluate(prediction, {re.metricName:'mse'})
## 21.920756856226326
```

可以看出，R^2 远超 50%，有较强的预测能力。

12.2 逻辑回归模型

考虑二分类问题，其输出标记 $y\in\{0,1\}$，而线性回归模型产生的预测值 $z=\boldsymbol{w}^{\mathrm{T}}\boldsymbol{x}_i+b$ 是实值，需要找到一个连续可微的函数将实值 z 转换为 0/1 值。对数概率函数正是这样一个函数：

$$y=\frac{1}{1+e^{-z}}=\frac{1}{1+e^{-(w^{T}x+b)}}$$

如图 12-2 所示。

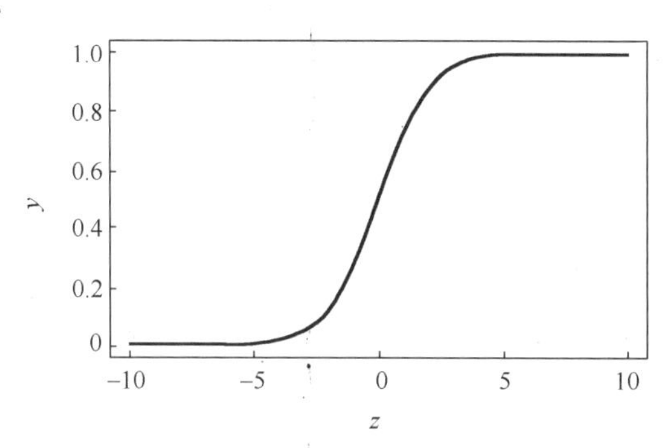

图 12-2　逻辑回归模型图

逻辑回归一般仅用于分类问题，而不用于回归问题。

调用 pysprk.ml.regression 程序包中的 LogisticRegression( )构造函数创建逻辑回归模型，得到超参数列表和每个超参数的默认值。

```
lr = LogisticRegression( )
print(lr.explainParams( ))
aggregationDepth: suggested depth for treeAggregate (>= 2). (default: 2)
elasticNetParam: the ElasticNet mixing parameter, in range [0, 1]. For alpha = 0, the penalty is an L2 penalty. For alpha = 1, it is an L1 penalty. (default: 0.0)
family: The name of family which is a description of the label distribution to be used in the model. Supported options: auto, binomial, multinomial (default: auto)
featuresCol: features column name. (default: features)
fitIntercept: whether to fit an intercept term. (default: True)
labelCol: label column name. (default: label)
lowerBoundsOnCoefficients: The lower bounds on coefficients if fitting under bound constrained optimization. The bound matrix must be compatible with the shape (1, number of features) for binomial regression, or (number of classes, number of features) for multinomial regression. (undefined)
lowerBoundsOnIntercepts: The lower bounds on intercepts if fitting under bound constrained optimization. The bounds vector size must beequal with 1 for binomial regression, or the number oflasses for multinomial regression. (undefined)
maxIter: max number of iterations (>= 0). (default: 100)
predictionCol: prediction column name. (default: prediction)
probabilityCol: Column name for predicted class conditional probabilities. Note: Not all models output well-calibrated probability estimates! These probabilities should be treated as confidences, not precise probabilities. (default: probability)
rawPredictionCol: raw prediction (a.k.a. confidence) column name. (default: rawPrediction)
regParam: regularization parameter (>= 0). (default: 0.0)
standardization: whether to standardize the training features before fitting the model. (default: True)
threshold: Threshold in binary classification prediction, in range [0, 1]. If threshold and thresholds are both set, they must match.e.g. if threshold is p, then thresholds must be equal to [1-p, p]. (default: 0.5)
thresholds: Thresholds in multi-class classification to adjust the probability of predicting each class. Array must have length equal to the number of classes, with values > 0, excepting that at most one value may be 0. The class with largest value p/t is predicted, where p is the original probability of that class and t is the class's threshold. (undefined)
```

```
tol: the convergence tolerance for iterative algorithms (>= 0). (default: 1e-06)
upperBoundsOnCoefficients: The upper bounds on coefficients if fitting under bound constrained optimization. The bound matrix must be compatible with the shape (1, number of features) for binomial regression, or (number of classes, number of features) for multinomial regression. (undefined)
upperBoundsOnIntercepts: The upper bounds on intercepts if fitting under bound constrained optimization. The bound vector size must be equal with 1 for binomial regression, or the number of classes for multinomial regression. (undefined)
weightCol: weight column name. If this is not set or empty, we treat all instance weights as 1.0. (undefined)
```

可以看出，有多个可调优的超参数，最常用的包括：

- 参数 regParam：规则化参数；
- 参数 standardization：是否归一化自变量，默认为是。

1．超参数调优

创建网格搜索超参数调优器，添加需要调优的超参数及尝试的数值，这里调优的超参数为：

- 规则化参数（regParam）：0.1、0.2 和 0.5；
- 是否归一化（standardization）：True 和 False。

```
grid = ParamGridBuilder( ).addGrid(lr.regParam,[0.1,0.2,0.5]) \
  .addGrid(lr.standardization,[True,False]).build( )
```

做 3 折交叉验证，得到最佳超参数组合是规则化参数（regParam）为 0.5 且是否归一化（standardization）为否。

```
cv = CrossValidator(estimator = lr, evaluator = BinaryClassification Evaluator( ),
  estimatorParamMaps = grid, numFolds = 3)
cvm = cv.fit(spirals)
print "regParam = ", cvm.bestModel._java_obj.parent( ).getRegParam( )
## regParam = 0.5
print "standardization = ", cvm.bestModel._java_obj.parent( ).get Standardization( )
## standardization = False
```

2．模型评估

调用逻辑回归模型的 transform()函数预测类别标签和其概率，这里为了突出模型，并没有分割训练集和测试集。

```
prediction = cvm.bestModel.transform(spirals)
prediction.show(5)
+--------------------+------+----------------------+----------------------+------------+
|          features  |label |    rawPrediction     |      probability     |prediction  |
+--------------------+------+----------------------+----------------------+------------+
| [0.333333333,0.0]  |   1.0|[0.00230019596917...|[0.50057504873874...|         0.0|
|[0.343933096,0.04...|   1.0|[0.00596116841877...|[0.50149028769151...|         0.0|
|[0.348689938,0.08...|   1.0|[0.00979904962986...|[0.50244974280519...|         0.0|
|[0.347116555,0.13...|   1.0|[0.01374397450669...|[0.50343593954033...|         0.0|
|[0.338838583,0.18...|   1.0|[0.01772026452935...|[0.50442995021304...|         0.0|
+--------------------+------+----------------------+----------------------+------------+
```

计算真阳性（tp）、假阳性（fp）、真阴性（tn）和假阴性（fn）样本数，并进一步计算模型精度、查准率和查全率。

```
tp = float(prediction.filter("prediction = 1.0 AND label = 1").count())
fp = float(prediction.filter("prediction = 1.0 AND label = 0").count())
tn = float(prediction.filter("prediction = 0.0 AND label = 0").count())
fn = float(prediction.filter("prediction = 0.0 AND label = 1").count())
accuracy = (tp + tn) / (tp + fp + tn + fn)
print(accuracy)
## 0.5
precision = tp / (tp + fp)
print(precision)
## 0.5
recall = tp / (tp + fn)
print(recall)
## 0.5
```

计算 ROC 曲线下面积。

```
bce = BinaryClassificationEvaluator(metricName = 'areaUnderROC')
bce.evaluate(prediction)
## 0.5720000000000002
```

可以看出，精度仅 50%，与随机猜测一样，毫无预测能力。

3．决策边界

画出逻辑回归模型的决策边界，如图 12-3 所示。

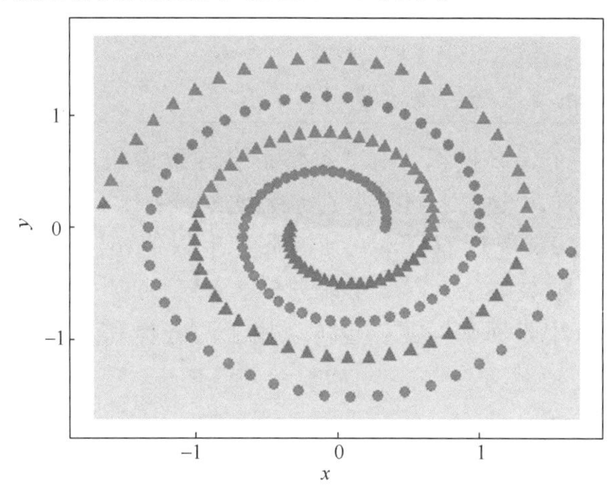

图 12-3　逻辑回归模型的决策边界（彩色图片见插页）

可以看出，逻辑回归模型的决策边界是一个超平面，无法学习出双螺旋结构这类高度非线性的决策边界。

12.3　决策树模型

一棵决策树包含一个根节点、若干个内部节点和若干个叶节点；叶节点对应于决策结果，其他每个节点则对应一个属性测试，每个节点包含的样本集合根据属性测试的结果被划分到子节点中；根节点包含样本全集。从根节点到每个叶节点的路径对应了一个判定测试序列。决策树模型的目的是产生一棵泛化能力强（即处理未见示例能力强）的决策树，其基本流程遵循简单且直观的“分而治之”策略。

输入： 训练集 $D=\{(\boldsymbol{x}_1,y_1),(\boldsymbol{x}_2,y_2),\cdots,(\boldsymbol{x}_m,y_m)\}$

属性集 $A=\{a_1,a_2,\cdots,a_d\}$

过程： 函数 TreeGenerate(D,A)

1: 生成节点 node
2: **if** D 中样本全属于同一类别 C **then**
3: 　将 node 标记为 C 类叶节点; **return**
4: **end if**
5: **if** $A=\varnothing$ **or** D 中样本在 A 上取值相同 **then**
6: 　将 node 标记为叶节点，其类别标记为 D 中样本数量最多的类; **return**
7: end if
8: 从 A 中选择最优划分属性 a_*
9: **for** a_* 的每一个值 a_*^v **do**
10: 　为 node 生成一个分支；令 D_v 表示 D 中在 a_* 上取值为 a_*^v 的样本子集
11: 　**if** D_v 为空 **then**
12: 　　将分支节点标记为叶节点，其类别标记为 D 中样本最多的类; **return**
13: 　**else**
14: 　　以 TreeGenerate($D,A\setminus\{a_*\}$) 为分支节点
15: 　**end if**
16: end for

输出： 以 node 为根节点的一棵决策树

决策树的生成是一个递归过程，有 3 种情形会导致递归返回：

- 当前节点包含的样本全属于同一类别，无须划分；
- 当前属性集为空，或是所有样本在所有属性上取值相同，无法划分；
- 当前节点包含的样本集合为空，不能划分。

决策树模型既可以用于回归问题，也可以用于两分类和多分类问题。

12.3.1 回归问题

调用 pyspark.ml.classification 程序包中的 DecisionTreeRegressor( )构造函数创建决策树回归器，得到超参数列表和每个超参数的默认值。

```
dtr = DecisionTreeRegressor( )
print(dtr.explainParams( ))
cacheNodeIds: If false, the algorithm will pass trees to executors to match instances with nodes. If true, the algorithm will cache node IDs for each instance. Caching can speed up training of deeper trees. Users can set how often should the cache be checkpointed or disable it by setting checkpointInterval. (default: False)
checkpointInterval: set checkpoint interval (>= 1) or disable checkpoint (-1). E.g. 10 means that the cache will get checkpointed every 10 iterations. Note: this setting will be ignored if the checkpoint directory is not set in the SparkContext. (default: 10)
featuresCol: features column name. (default: features)
impurity: Criterion used for information gain calculation (case-insensitive). Supported options: variance (default: variance)
labelCol: label column name. (default: label)
```

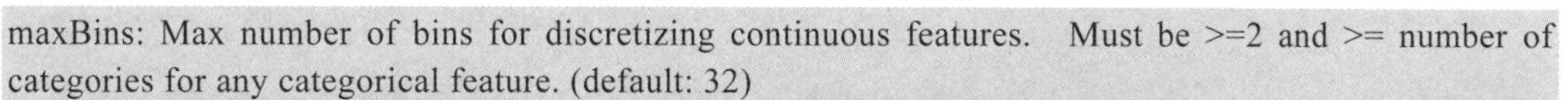

```
maxBins: Max number of bins for discretizing continuous features.  Must be >=2 and >= number of categories for any categorical feature. (default: 32)
maxDepth: Maximum depth of the tree. (>= 0) E.g., depth 0 means 1 leaf node; depth 1 means 1 internal node + 2 leaf nodes. (default: 5)
maxMemoryInMB: Maximum memory in MB allocated to histogram aggregation. If too small, then 1 node will be split per iteration, and its aggregates may exceed this size. (default: 256)
minInfoGain: Minimum information gain for a split to be considered at a tree node. (default: 0.0)
minInstancesPerNode: Minimum number of instances each child must have after split. If a split causes the left or right child to have fewer than minInstancesPerNode, the split will be discarded as invalid. Should be >= 1. (default: 1)
predictionCol: prediction column name. (default: prediction)
seed: random seed. (default: -2808853809871465425)
varianceCol: column name for the biased sample variance of prediction. (undefined)
```

可以看出，有多个可调优的超参数，最常用的包括：

- 参数 maxDepth：树的最大深度；
- 参数 minInfoGain：最小信息增益。

1．超参数调优

创建网格搜索超参数调优器，添加需要调优的超参数及尝试的数值，这里调优的超参数为：

- 最大深度（maxDepth）：3、4 和 5；
- 最小信息增益（minInfoGain）：0.001 和 0.005。

```
grid =  ParamGridBuilder().addGrid(dtr.maxDepth,[3,4,5]) \
  .addGrid(dtr.minInfoGain,[0.001,0.005]).build()
```

做 3 折交叉验证，得到最佳超参数组合是最大深度（maxDepth）为 4 且最小信息增益（minInfoGain）为 0.001。

```
cv = CrossValidator(estimator = dtr,evaluator = RegressionEvaluator(),
  estimatorParamMaps = grid, numFolds = 3)
cvm = cv.fit(boston)
print "maxDepth = ", cvm.bestModel._java_obj.parent().getMaxDepth()
## maxDepth =  4
print "minInfoGain = ", cvm.bestModel._java_obj.parent(). getMinInfoGain()
## minInfoGain =  0.001
```

2．模型评估

调用决策树回归模型的 transform()函数预测数值标签，这里为了突出模型，并没有分割训练集和测试集。

```
prediction = cvm.bestModel.transform(boston)
prediction.show(5)
+--------------------+-----+------------------+
|       features     |label|    prediction    |
+--------------------+-----+------------------+
|  [0.00632,18.0,2.3...| 24.0| 27.51800000000002 |
|  [0.02731,0.0,7.07...| 21.6|21.779999999999998|
|  [0.02729,0.0,7.07...| 34.7|32.406976744186046|
```

```
|    [0.03237,0.0,2.18...| 33.4|32.406976744186046|
|    [0.06905,0.0,2.18...| 36.2|32.406976744186046|
+-------------------------+-----+----------------------+
```

计算模型性能指标。

```
re = RegressionEvaluator( )
re.evaluate(prediction, {re.metricName:'r2'})
## 0.8754196511956004
re.evaluate(prediction, {re.metricName:'mse'})
## 10.517017751847694
```

可以看出，R^2 远超 50%，有较强的预测能力。

12.3.2　二分类问题

调用 pyspark.ml.classification 程序包中的 DecisionTreeClassifier( )构造函数创建决策树模型，得到超参数列表和每个超参数的默认值。

```
dtc = DecisionTreeClassifier()
print(dtc.explainParams( ))
cacheNodeIds: If false, the algorithm will pass trees to executors to match instances with nodes. If true, the algorithm will cache node IDs for each instance. Caching can speed up training of deeper trees. Users can set how often should the cache be checkpointed or disable it by setting checkpointInterval. (default: False)
checkpointInterval: set checkpoint interval (>= 1) or disable checkpoint (-1). E.g. 10 means that the cache will get checkpointed every 10 iterations. Note: this setting will be ignored if the checkpoint directory is not set in the SparkContext. (default: 10)
featuresCol: features column name. (default: features)
impurity: Criterion used for information gain calculation (case-insensitive). Supported options: entropy, gini (default: gini)
labelCol: label column name. (default: label)
maxBins: Max number of bins for discretizing continuous features.  Must be >=2 and >= number of categories for any categorical feature. (default: 32)
maxDepth: Maximum depth of the tree. (>= 0) E.g., depth 0 means 1 leaf node; depth 1 means 1 internal node + 2 leaf nodes. (default: 5)
maxMemoryInMB: Maximum memory in MB allocated to histogram aggregation. If too small, then 1 node will be split per iteration, and its aggregates may exceed this size. (default: 256)
minInfoGain: Minimum information gain for a split to be considered at a tree node. (default: 0.0)
minInstancesPerNode: Minimum number of instances each child must have after split. If a split causes the left or right child to have fewer than minInstancesPerNode, the split will be discarded as invalid. Should be >= 1. (default: 1)
predictionCol: prediction column name. (default: prediction)
probabilityCol: Column name for predicted class conditional probabilities. Note: Not all models output well-calibrated probability estimates! These probabilities should be treated as confidences, not precise probabilities. (default: probability)
rawPredictionCol: raw prediction (a.k.a. confidence) column name. (default: rawPrediction)
seed: random seed. (default: 6174923023070228847)
```

可以看出，有多个可调优的超参数，最常用的包括：

- 参数 maxDepth：树的最大深度；
- 参数 minInfoGain：最小信息增益。

1．超参数调优

创建网格搜索超参数调优器，添加需要调优的超参数及尝试的数值，这里调优的超参数为：

- 最大深度（maxDepth）：3、4 和 5；
- 最小信息增益（minInfoGain）：0.001 和 0.005。

```
grid =  ParamGridBuilder().addGrid(dtc.maxDepth,[3,4,5]) \
  .addGrid(dtc.minInfoGain,[0.001,0.005]).build()
```

做 3 折交叉验证器，得到最佳超参数组合是最大深度（maxDepth）为 5 且最小信息增益（minInfoGain）为 0.001。

```
cv = CrossValidator(estimator = dtc,evaluator = BinaryClassifi cationEvaluator(),
  estimatorParamMaps = grid, numFolds = 3)
cvm = cv.fit(spirals)
print "maxDepth = ", cvm.bestModel._java_obj.parent().getMaxDepth()
## maxDepth =  5
print "minInfoGain = ", cvm.bestModel._java_obj.parent(). getMinInfoGain()
## minInfoGain =  0.001
```

2．模型评估

调用决策树分类模型的 transform()函数预测类别标签及其概率，这里为了突出模型，并没有分割训练集和测试集。

```
prediction = cvm.bestModel.transform(spirals)
prediction.show(5)
+----------------------+------+--------------+------------------------+-----------+
|       features       |label |rawPrediction|       probability      |prediction|
+----------------------+------+--------------+------------------------+-----------+
| [0.333333333,0.0]  |  1.0|    [30.0,8.0]|[0.78947368421052...|        0.0|
|[0.343933096,0.04...|  1.0|    [7.0,29.0]|[0.19444444444444...|        1.0|
|[0.348689938,0.08...|  1.0|    [7.0,29.0]|[0.19444444444444...|        1.0|
|[0.347116555,0.13...|  1.0|    [7.0,29.0]|[0.19444444444444...|        1.0|
|[0.338838583,0.18...|  1.0|    [7.0,29.0]|[0.19444444444444...|        1.0|
+----------------------+------+--------------+------------------------+-----------+
```

计算真阳性（tp）、假阳性（fp）、真阴性（tn）和假阴性（fn）样本数，并进一步计算模型精度、查准率和查全率。

```
tp = float(prediction.filter("prediction = 1.0 AND label = 1").count())
fp = float(prediction.filter("prediction = 1.0 AND label = 0").count())
tn = float(prediction.filter("prediction = 0.0 AND label = 0").count())
fn = float(prediction.filter("prediction = 0.0 AND label = 1").count())
accuracy = (tp + tn) / (tp + fp + tn + fn)
```

```
print(accuracy)
## 0.775
precision = tp / (tp + fp)
print(precision)
## 0.777777777778
recall = tp / (tp + fn)
print(recall)
## 0.77
```

计算 ROC 曲线下面积。

```
bce = BinaryClassificationEvaluator(metricName = 'areaUnderROC')
bce.evaluate(prediction)
## 0.7211
```

可以看出，精度远超 50%，有较强的预测能力。

3．决策边界

画出决策树模型的决策边界，如图 12-4 所示。

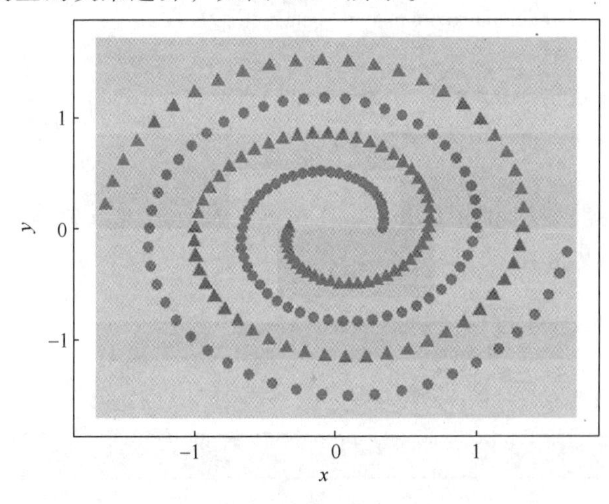

图 12-4　决策树模型的决策边界（彩色图片见插页）

可以看出，决策树模型的决策边界是平行于坐标轴的超平面组成的超曲面，可以部分学习出双螺旋结构这类高度非线性的决策边界。

12.3.3　多分类问题

调用 pyspark.ml.classification 程序包中的 DecisionTreeClassifier( )构造函数创建决策树分类器。

```
dtc = DecisionTreeClassifier( )
```

1．超参数调优

调用 pyspark.ml.tuning 程序包中的 ParamGridBuilder()构造函数创建网格搜索超参数调优器。

调用网格搜索超参数调优器的 addGrid( )函数添加需要调优的超参数及尝试的数

值，这里调优的超参数为

- 最大深度（maxDepth）：3、4 和 5；
- 最小信息增益（minInfoGain）：0.001 和 0.005。

```
grid =  ParamGridBuilder( ).addGrid(dtc.maxDepth,[3,4,5]) \
  .addGrid(dtc.minInfoGain,[0.001,0.005]).build( )
```

调用 pyspark.ml.tuning 程序包中的 CrossValidator()构造函数创建交叉验证器。

```
cv = CrossValidator(estimator = dtc,evaluator = MulticlassClassi ficationEvaluator( ),
  estimatorParamMaps = grid, numFolds = 3)
```

调用交叉验证器的 fit()函数做超参数调优，输入参数为数据框，返回训练后的交叉验证器模型。

使用交叉验证器模型的 bestModel 属性得到最优的机器学习模型，并使用其属性 _java_obj 中的相应函数 get<超参数名称>( )得到最佳超参数组合是最大深度（maxDepth）为 3 且最小信息增益（minInfoGain）为 0.001。

```
cvm = cv.fit(iris)
print "maxDepth = ", cvm.bestModel._java_obj.parent( ).getMaxDepth( )
## maxDepth =   4
print "minInfoGain = ", cvm.bestModel._java_obj.parent( ).getMinInfoGain( )
## minInfoGain =   0.001
```

2. 模型评估

调用决策树分类模型的 transform()函数预测类别标签及其概率，这里为了突出模型，并没有分割训练集和测试集。

```
prediction = cvm.bestModel.transform(iris)
prediction.show(5)
+-----------------+-----+------------------+-----------------+------------+
|    features     |label| rawPrediction|  probability |prediction|
+-----------------+-----+------------------+-----------------+------------+
|[5.1,3.5,1.4,0.2]|  0.0|[50.0,0.0,0.0]  |[1.0,0.0,0.0]   |        0.0|
|[4.9,3.0,1.4,0.2]|  0.0|[50.0,0.0,0.0]  |[1.0,0.0,0.0]   |        0.0|
|[4.7,3.2,1.3,0.2]|  0.0|[50.0,0.0,0.0]  |[1.0,0.0,0.0]   |        0.0|
|[4.6,3.1,1.5,0.2]|  0.0|[50.0,0.0,0.0]  |[1.0,0.0,0.0]   |        0.0|
|[5.0,3.6,1.4,0.2]|  0.0|[50.0,0.0,0.0]  |[1.0,0.0,0.0]   |        0.0|
+-----------------+-----+------------------+-----------------+------------+
```

计算模型精度。

```
accuracy = float(prediction.filter("prediction = label").count( )) /   prediction.count( )
print(accuracy)
## 0.993333333333
```

可以看出，精度远超 50%，有较强的预测能力。

3. 决策边界

画出决策树模型在 2 个自变量上的决策边界，如图 12-5 所示。

可以看出，决策树模型的决策边界是平行于坐标轴的超平面组成的超曲面。

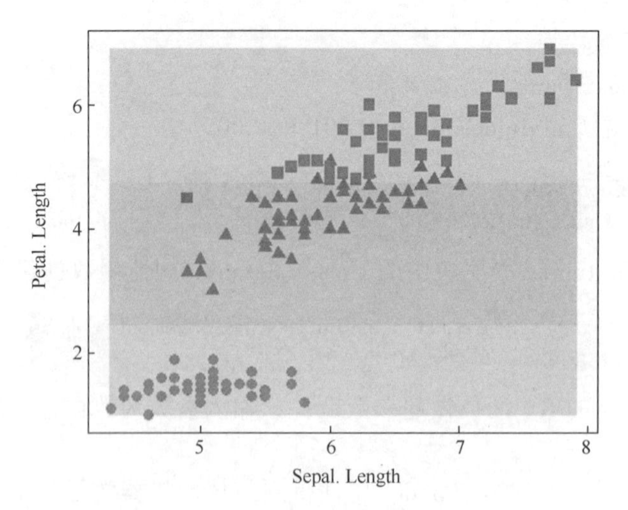

图 12-5　决策树模型在 2 个自变量上的决策边界（彩色图片见插页）

12.4　随机森林模型

随机森林属于集成学习的一种，由一组决策树组成，实现简单，性能强大。对于包含 m 个样本的训练集，进行 m 次重采样得到采样集，初始训练集中有的样本多次出现，有的则从未出现，约有 63.2%的样本在采样集中，用于每个决策树的训练。对其中每个决策树的每个节点，先随机选择一个包含 k 个属性的子集，再从中选择一个最优属性用于划分。

随机森林模型既可以用于回归问题，也可以用于两分类和多分类问题。

12.4.1　回归问题

调用 pyspark.ml.regression 程序包中的 RandomForestRegressor()构造函数创建随机森林回归器，得到超参数列表和每个超参数的默认值。

```
rfr = RandomForestRegressor( )
print(rfr.explainParams( ))
cacheNodeIds: If false, the algorithm will pass trees to executors to match instances with nodes. If true, the algorithm will cache node IDs for each instance. Caching can speed up training of deeper trees. Users can set how often should the cache be checkpointed or disable it by setting checkpointInterval. (default: False)
checkpointInterval: set checkpoint interval (>= 1) or disable checkpoint (-1). E.g. 10 means that the cache will get checkpointed every 10 iterations. Note: this setting will be ignored if the checkpoint directory is not set in the SparkContext. (default: 10)
featureSubsetStrategy: The number of features to consider for splits at each tree node. Supported options: 'auto' (choose automatically for task: If numTrees == 1, set to 'all'. If numTrees > 1 (forest), set to 'sqrt' for classification and to 'onethird' for regression), 'all' (use all features), 'onethird' (use 1/3 of the features), 'sqrt' (use sqrt(number of features)), 'log2' (use log2(number of features)), 'n' (when n is in the range (0, 1.0], use n * number of features. When n is in the range (1, number of features), use n features). default = 'auto' (default: auto)
featuresCol: features column name. (default: features)
impurity: Criterion used for information gain calculation (case-insensitive). Supported options: variance (default: variance)
```

```
labelCol: label column name. (default: label)
maxBins: Max number of bins for discretizing continuous features.  Must be >=2 and >= number of categories for any categorical feature. (default: 32)
maxDepth: Maximum depth of the tree. (>= 0) E.g., depth 0 means 1 leaf node; depth 1 means 1 internal node + 2 leaf nodes. (default: 5)
maxMemoryInMB: Maximum memory in MB allocated to histogram aggregation. If too small, then 1 node will be split per iteration, and its aggregates may exceed this size. (default: 256)
minInfoGain: Minimum information gain for a split to be considered at a tree node. (default: 0.0)
minInstancesPerNode: Minimum number of instances each child must have after split. If a split causes the left or right child to have fewer than minInstancesPerNode, the split will be discarded as invalid. Should be >= 1. (default: 1)
numTrees: Number of trees to train (>= 1). (default: 20)
predictionCol: prediction column name. (default: prediction)
seed: random seed. (default: -5851613654371098793)
subsamplingRate: Fraction of the training data used for learning each decision tree, in range (0, 1]. (default: 1.0)
```

可以看出，有多个可调优的超参数，最常用的包括：

- 参数 NumTrees：树的数量；
- 参数 FeatureSubsetStrategy：划分节点时随机抽取候选特征数量的策略，常用选项包括 auto（表示自动策略）、all（表示特征总数）、onethird（表示特征总数的三分之一）、sqrt（表示特征总数的平方根）、log2（表示特征总数的对数）；
- 参数 MinInstancesPerNode：树的叶节点包含的最小样本数。

1. 超参数调优

创建网格搜索超参数调优器，添加需要调优的超参数及尝试的数值，这里调优的超参数为：

- 树的数量（numTrees）：10、20 和 50；
- 划分节点时随机抽取候选特征数量的策略（featureSubsetStrategy）：all、onethird 和 sqrt。

```
grid = ParamGridBuilder( ).addGrid(rfr.numTrees,[10,20,50]) \
  .addGrid(rfr.featureSubsetStrategy,['all', 'onethird', 'sqrt']).build( )
```

做 3 折交叉验证器，得到最佳超参数组合是树的数量（numTrees）为 50 且划分节点时随机抽取候选特征数量的策略（featureSubsetStrategy）为特征总数的三分之一（onethird）。

```
cv = CrossValidator(estimator = rfr, evaluator = RegressionEvaluator( ), estimatorParamMaps = grid, numFolds = 3)
cvm = cv.fit(boston)
print "numTrees = ", cvm.bestModel._java_obj.parent( ).getNumTrees( )
## numTrees =   50
print "featureSubsetStrategyr = ", cvm.bestModel._java_obj. parent( ).getFeatureSubsetStrategy( )
## featureSubsetStrategyr = all
```

2. 模型评估

调用随机森林回归模型的 transform()函数预测数值标签，这里为了突出模型，并没有分割训练集和测试集。

```
prediction = cvm.bestModel.transform(boston)
prediction.show(5)
+--------------------+-----+--------------------+
|        features    |label|      prediction    |
+--------------------+-----+--------------------+
|[0.00632,18.0,2.3... | 24.0|27.167873018830548|
|[0.02731,0.0,7.07... | 21.6| 22.74863338400353 |
|[0.02729,0.0,7.07... | 34.7|   33.8644081846892|
|[0.03237,0.0,2.18... | 33.4| 33.13745015157199 |
|[0.06905,0.0,2.18... | 36.2|   33.8114149701551|
+--------------------+-----+--------------------+
```

计算模型性能指标。

```
re = RegressionEvaluator( )
re.evaluate(prediction, {re.metricName:'r2'})
## 0.9345231563922636
re.evaluate(prediction, {re.metricName:'mse'})
## 5.5275260758717675
```

可以看出，R^2 远超 50%，有较强的预测能力。

12.4.2 二分类问题

调用 pyspark.ml.regression 程序包中的 RandomForestClassifier( )构造函数创建随机森林分类器，得到超参数列表和每个超参数的默认值。

```
rfc = RandomForestClassifier( )
print(rfc.explainParams( ))
cacheNodeIds: If false, the algorithm will pass trees to executors to match instances with nodes. If true, the algorithm will cache node IDs for each instance. Caching can speed up training of deeper trees. Users can set how often should the cache be checkpointed or disable it by setting checkpointInterval. (default: False)
checkpointInterval: set checkpoint interval (>= 1) or disable checkpoint (-1). E.g. 10 means that the cache will get checkpointed every 10 iterations. Note: this setting will be ignored if the checkpoint directory is not set in the SparkContext. (default: 10)
featureSubsetStrategy: The number of features to consider for splits at each tree node. Supported options: 'auto' (choose automatically for task: If numTrees == 1, set to 'all'. If numTrees > 1 (forest), set to 'sqrt' for classification and to 'onethird' for regression), 'all' (use all features), 'onethird' (use 1/3 of the features), 'sqrt' (use sqrt(number of features)), 'log2' (use log2(number of features)), 'n' (when n is in the range (0, 1.0], use n * number of features. When n is in the range (1, number of features), use n features). default = 'auto' (default: auto)
featuresCol: features column name. (default: features)
impurity: Criterion used for information gain calculation (case-insensitive). Supported options: entropy, gini (default: gini)
labelCol: label column name. (default: label)
maxBins: Max number of bins for discretizing continuous features.  Must be >=2 and >= number of categories for any categorical feature. (default: 32)
```

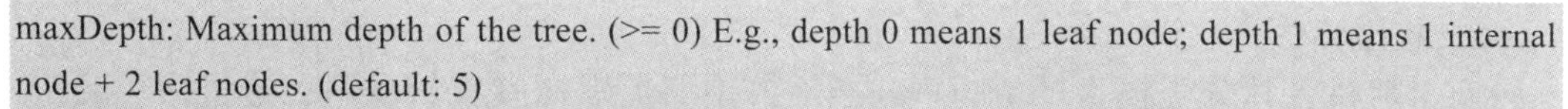

```
maxDepth: Maximum depth of the tree. (>= 0) E.g., depth 0 means 1 leaf node; depth 1 means 1 internal node + 2 leaf nodes. (default: 5)
maxMemoryInMB: Maximum memory in MB allocated to histogram aggregation. If too small, then 1 node will be split per iteration, and its aggregates may exceed this size. (default: 256)
minInfoGain: Minimum information gain for a split to be considered at a tree node. (default: 0.0)
minInstancesPerNode: Minimum number of instances each child must have after split. If a split causes the left or right child to have fewer than minInstancesPerNode, the split will be discarded as invalid. Should be >= 1. (default: 1)
numTrees: Number of trees to train (>= 1). (default: 20)
predictionCol: prediction column name. (default: prediction)
probabilityCol: Column name for predicted class conditional probabilities. Note: Not all models output well-calibrated probability estimates! These probabilities should be treated as confidences, not precise probabilities. (default: probability)
rawPredictionCol: raw prediction (a.k.a. confidence) column name. (default: rawPrediction)
seed: random seed. (default: -4140900678877021401)
subsamplingRate: Fraction of the training data used for learning each decision tree, in range (0, 1]. (default: 1.0)
```

可以看出，有多个可调优的超参数，最常用的包括：

- 参数 NumTrees：树的数量；
- 参数 FeatureSubsetStrategy：划分节点时随机抽取候选特征数量的策略，常用选项包括 auto（表示自动策略）、all（表示特征总数）、onethird（表示特征总数的三分之一）、sqrt（表示特征总数的平方根）、log2（表示特征总数的对数）；
- 参数 MinInstancesPerNode：树的叶节点包含的最小样本数。

1. 超参数调优

创建网格搜索超参数调优器，添加需要调优的超参数及尝试的数值，这里调优的超参数为：

- 树的数量（numTrees）：10、20 和 50；
- 划分节点时随机抽取候选特征数量的策略（featureSubsetStrategy）：all、onethird 和 sqrt。

```
grid = ParamGridBuilder().addGrid(rfc.numTrees,[10,20,50]) \
   .addGrid(rfc.featureSubsetStrategy,['all','onethird','sqrt']).build()
```

做 3 折交叉验证器，得到最佳超参数组合是树的数量（numTrees）为 20 且划分节点时随机抽取候选特征数量的策略（featureSubsetStrategy）为特征总数（all）。

```
cv = CrossValidator(estimator = rfc, evaluator = BinaryClassif icationEvaluator(),
   estimatorParamMaps = grid, numFolds = 3)
cvm = cv.fit(spirals)
print "numTrees = ", cvm.bestModel._java_obj.parent().getNumTrees()
## numTrees = 20
print "featureSubsetStrategyr = ", cvm.bestModel._java_obj. parent().getFeatureSubsetStrategy()
## featureSubsetStrategyr = all
```

2．模型评估

调用随机森林分类模型的 transform()函数预测类别标签及其概率，这里为了突出模型，并没有分割训练集和测试集。

```
prediction = cvm.bestModel.transform(spirals)
prediction.show(5)
+-----------------------+------+-----------------------+----------------------------+-------------+
|        features       |label |    rawPrediction      |                probability | prediction  |
+-----------------------+------+-----------------------+----------------------------+-------------+
| [0.333333333,0.0]    |   1.0|[11.6844510521130... |[0.58422255260565...     |          0.0|
|[0.343933096,0.04... |   1.0|[6.84720443279810...|[0.34236022163990...     |          1.0|
|[0.348689938,0.08... |   1.0|[7.4397970253907,... |[0.37198985126953...     |          1.0|
|[0.347116555,0.13... |   1.0|[7.4397970253907,... |[0.37198985126953...     |          1.0|
|[0.338838583,0.18... |   1.0|[6.84720443279810...|[0.34236022163990...     |          1.0|
+-----------------------+------+-----------------------+----------------------------+-------------+
```

计算真阳性（tp）、假阳性（fp）、真阴性（tn）和假阴性（fn）样本数，并进一步计算模型精度、查准率和查全率。

```
tp = float(prediction.filter("prediction = 1.0 AND label = 1").count( ))
fp = float(prediction.filter("prediction = 1.0 AND label = 0").count( ))
tn = float(prediction.filter("prediction = 0.0 AND label = 0").count( ))
fn = float(prediction.filter("prediction = 0.0 AND label = 1").count( ))
accuracy = (tp + tn) / (tp + fp + tn + fn)
print(accuracy)
## 0.92
precision = tp / (tp + fp)
print(precision)
## 0.911764705882
recall = tp / (tp + fn)
print(recall)
## 0.93
```

计算 ROC 曲线下面积。

```
bce = BinaryClassificationEvaluator(metricName = 'areaUnderROC')
bce.evaluate(prediction)
# 0.9758500000000002
```

可以看出，精度远超 50%，有较强的预测能力。

3．决策边界

画出随机森林模型的决策边界，如图 12-6 所示。

可以看出，随机森林模型的决策边界是平行于坐标轴的超平面组成的超曲面，可以学习出双螺旋结构这类高度非线性的决策边界。

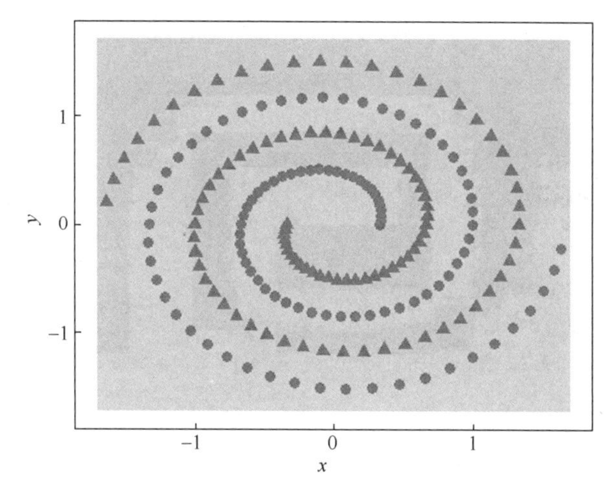

图 12-6 随机森林模型的决策边界（彩色图片见插页）

12.4.3 多分类问题

调用 pyspark.ml.regression 程序包中的 RandomForestClassifier()构造函数创建随机森林分类器。

```
rfc = RandomForestClassifier()
```

1．超参数调优

创建网格搜索超参数调优器，添加需要调优的超参数及尝试的数值，这里调优的超参数为：

- 树的数量（numTrees）：10、20 和 50；
- 划分节点时随机抽取候选特征数量的策略（featureSubsetStrategy）：all、onethird 和 sqrt。

```
grid = ParamGridBuilder().addGrid(rfc.numTrees,[10,20,50]).add Grid(rfc.featureSubsetStrategy,
['all', 'onethird','sqrt']).build()
```

做 3 折交叉验证器，得到最佳超参数组合是树的数量（numTrees）为 20 且划分节点时随机抽取候选特征数量的策略（featureSubsetStrategy）为特征总数（all）。

```
cv = CrossValidator(estimator = rfc, evaluator = MulticlassClassificationEvaluator(),
estimatorParamMaps = grid, numFolds = 3)
cvm = cv.fit(iris)
print "numTrees = ", cvm.bestModel._java_obj.parent().getNumTrees()
## numTrees = 20
print "featureSubsetStrategyr = ", cvm.bestModel._java_obj. parent().getFeatureSubsetStrategy()
## featureSubsetStrategyr = onethird
```

2．模型评估

调用随机森林分类模型的 transform()函数预测类别标签及其概率，这里为了突出模型，并没有分割训练集和测试集。

```
prediction = cvm.bestModel.transform(iris)
```

```
prediction.show(5)
+-----------------------+-----+-----------------+-------------+-----------+
|               features|label| rawPrediction |probability|prediction|
+-----------------------+-----+-----------------+-------------+-----------+
|[5.1,3.5,1.4,0.2]      |  0.0| [20.0,0.0,0.0]  | [1.0,0.0,0.0]|        0.0|
|[4.9,3.0,1.4,0.2]      |  0.0| [20.0,0.0,0.0]  | [1.0,0.0,0.0]|        0.0|
|[4.7,3.2,1.3,0.2]      |  0.0| [20.0,0.0,0.0]  | [1.0,0.0,0.0]|        0.0|
|[4.6,3.1,1.5,0.2]      |  0.0| [20.0,0.0,0.0]  | [1.0,0.0,0.0]|        0.0|
|[5.0,3.6,1.4,0.2]      |  0.0| [20.0,0.0,0.0]  | [1.0,0.0,0.0]|        0.0|
+-----------------------+-----+-----------------+-------------+-----------+
```

计算模型精度。

```
accuracy = float(prediction.filter("prediction = label").count( )) /   prediction.count( )
print(accuracy)
## 1.0
```

可以看出，精度远超 50%，有较强的预测能力。

3．决策边界

画出随机森林模型在 2 个自变量上的决策边界，如图 12-7 所示。

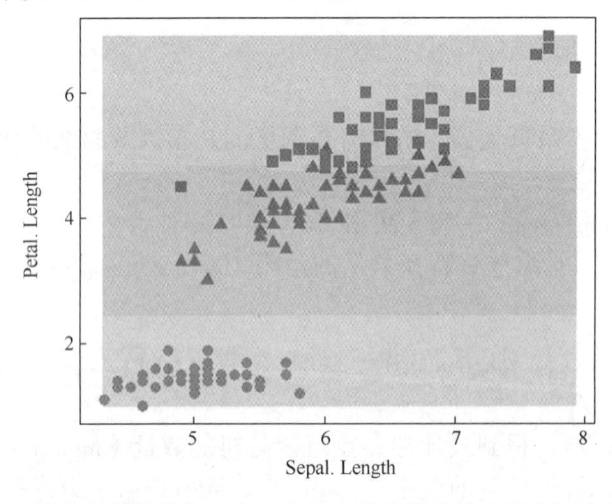

图 12-7　随机森林模型在 2 个自变量上的决策边界（彩色图片见插页）

可以看出，随机森林模型的决策边界是平行于坐标轴的超平面组成的超曲面。

12.5　神经网络

神经网络是由具有适应性的简单单元组成的广泛并行互连的网络，它的组织能够模拟生物神经系统对真实世界物体所作出的交互反应。

神经网络最基本的成分是 M-P 神经元（neuron）模型，神经元接收到来自 n 个其他神经元传递过来的输入信号（$x_1,x_2,\cdots,x_n$），通过带权重的连接（$w_1,w_2,\cdots,w_n$）进行传递，神经元接收到的总输入值将与神经元的阈值（θ）进行比较，然后通过激活函数处理以产生神经元的输出，如图 12-8 所示。

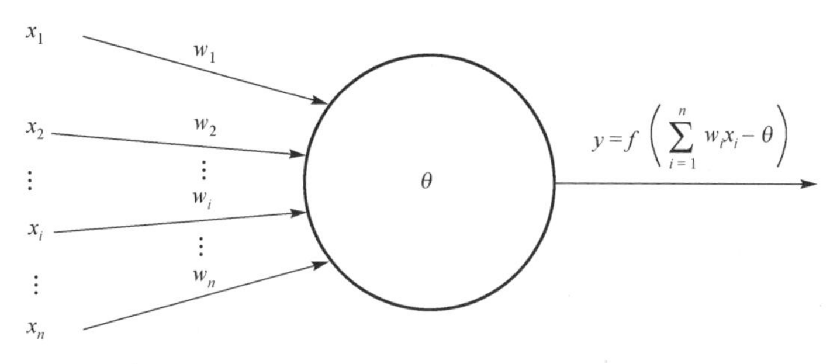

图 12-8 神经元接收过程图

常见的神经网络是如图 12-9 所示的层级结构，每层神经元与下一层神经元全互连，神经元之间不存在同层连接，也不存在跨层连接，这样的神经网络结构称为多层前馈神经网络。

图 12-9 所示为一个 2 层前馈神经网络，其中有 1 个隐藏层，最左侧为输入层，最右侧为输出层。

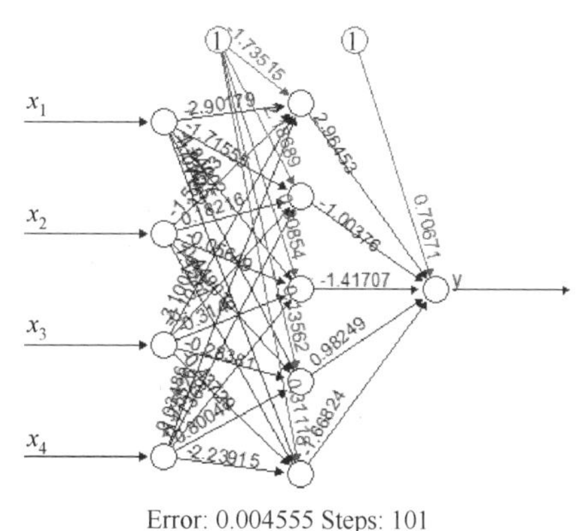

图 12-9 2 层前馈神经网络图

神经网络模型既可以用于回归问题，也可以用于两分类和多分类问题。

12.5.1 二分类问题

调用 pysprk.ml.classification 程序包中的 MultilayerPerceptronClassifier( )构造函数创建神经网络分类器，得到超参数列表和每个超参数的默认值。

```
mpc = MultilayerPerceptronClassifier( )
print(mpc.explainParams( ))
blockSize: Block size for stacking input data in matrices. Data is stacked within partitions. If block size is more than remaining data in a partition then it is adjusted to the size of this data. Recommended size is between 10 and 1000, default is 128. (default: 128)
featuresCol: features column name. (default: features)
initialWeights: The initial weights of the model. (undefined)
labelCol: label column name. (default: label)
layers: Sizes of layers from input layer to output layer E.g., Array(780, 100, 10) means 780 inputs, one hidden layer with 100 neurons and output layer of 10 neurons. (undefined)
```

```
maxIter: max number of iterations (>= 0). (default: 100)
predictionCol: prediction column name. (default: prediction)
probabilityCol: Column name for predicted class conditional probabilities. Note: Not all models output well-calibrated probability estimates! These probabilities should be treated as confidences, not precise probabilities. (default: probability)
rawPredictionCol: raw prediction (a.k.a. confidence) column name. (default: rawPrediction)
seed: random seed. (default: 4423497804127365397)
solver: The solver algorithm for optimization. Supported options: l-bfgs, gd. (default: l-bfgs)
stepSize: Step size to be used for each iteration of optimization (>= 0). (default: 0.03)
tol: the convergence tolerance for iterative algorithms (>= 0). (default: 1e-06)
```

可以看出，有多个可调优的超参数，最常用的包括：

- 参数 layers：从输入层到输出层的每一层神经元数量的列表，其中第一个元素的值必须与输入特征数一致，最后一个元素的值必须与类别数一致。
- 参数 maxIter：最大迭代次数；
- 参数 tol：收敛容忍度。

1．超参数调优

创建网格搜索超参数调优器，添加需要调优的超参数及尝试的数值，这里调优的超参数为

- 最大迭代次数（maxIter）：20、50 和 100；
- 每一层神经元数量（layers）：[2,20,20,2]和[2,50,50,2]。

```
grid = ParamGridBuilder( ).addGrid(mpc.maxIter,[20,50,100]) \
   .addGrid(mpc.layers,[[2,20,20,2],[2,50,50,2]]).build( )
```

做 3 折交叉验证器。

调用交叉验证器的 fit()函数做超参数调优，输入参数为数据框，返回训练后的交叉验证器模型，得到最佳超参数组合是最大迭代次数（maxIter）为 100 且每一层神经元数量（layers）为[2,50,50,2]。

```
cv = CrossValidator(estimator = mpc, evaluator = MulticlassClassi ficationEvaluator( ),
   estimatorParamMaps = grid, numFolds = 3)
cvm = cv.fit(spirals)
print "maxIter = ", cvm.bestModel._java_obj.parent( ).getMaxIter( )
## maxIter = 100
print "layers = ", cvm.bestModel.layers
## layers = [2,20,20,2]
```

2．模型评估

调用神经网络分类模型的 transform()函数预测数值标签，这里为了突出模型，并没有分割训练集和测试集。

```
prediction = cvm.bestModel.transform(spirals)
prediction.show(5)
+----------------------+------+-------------------------+-------------------------+-------------+
|        features      |label |     rawPrediction       |       probability       |prediction  |
+----------------------+------+-------------------------+-------------------------+-------------+
```

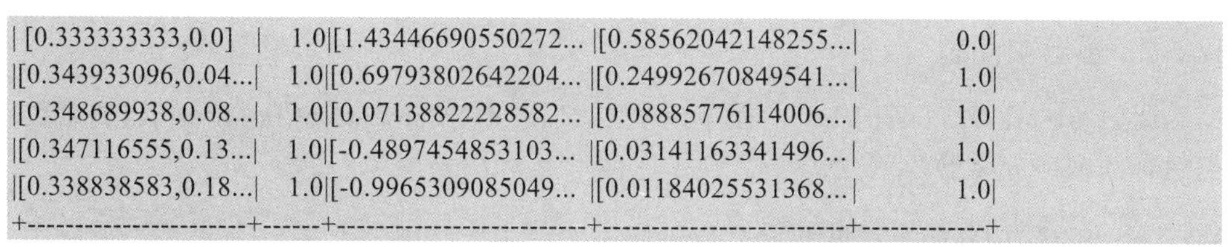

```
| [0.333333333,0.0]  |  1.0|[1.43446690550272...|[0.58562042148255...|          0.0|
|[0.343933096,0.04...|  1.0|[0.69793802642204...|[0.24992670849541...|          1.0|
|[0.348689938,0.08...|  1.0|[0.07138822228582...|[0.08885776114006...|          1.0|
|[0.347116555,0.13...|  1.0|[-0.4897454853103...|[0.03141163341496...|          1.0|
|[0.338838583,0.18...|  1.0|[-0.9965309085049...|[0.01184025531368...|          1.0|
+----------------------+------+-------------------------+------------------------+-------------+
```

计算真阳性（tp）、假阳性（fp）、真阴性（tn）和假阴性（fn）样本数，并进一步计算模型精度、查准率和查全率。

```
tp = float(prediction.filter("prediction = 1.0 AND label = 1").count())
fp = float(prediction.filter("prediction = 1.0 AND label = 0").count())
tn = float(prediction.filter("prediction = 0.0 AND label = 0").count())
fn = float(prediction.filter("prediction = 0.0 AND label = 1").count())
accuracy = (tp + tn) / (tp + fp + tn + fn)
print(accuracy)
## 0.88
precision = tp / (tp + fp)
print(precision)
## 0.913043478261
recall = tp / (tp + fn)
print(recall)
## 0.84
```

计算 ROC 曲线下面积。

```
bce = BinaryClassificationEvaluator(metricName = 'areaUnderROC')
bce.evaluate(prediction)
# 0. 9551000000000001
```

可以看出，精度远超 50%，有较强的预测能力。

3. 决策边界

画出神经网络模型的决策边界，如图 12-10 所示。

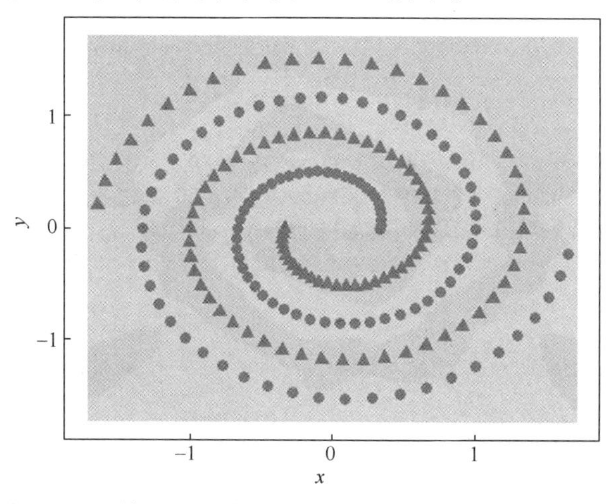

图 12-10　神经网络模型的决策边界（彩色图片见插页）

可以看出，神经网络模型的决策边界可以是一个不规则超曲面，可以学习出双螺旋结构这类高度非线性的决策边界。

12.5.2 多分类问题

调用 pysprk.ml.classification 程序包中的 MultilayerPerceptronClassifier( )构造函数创建神经网络分类器。

```
mpc = MultilayerPerceptronClassifier( )
```

1. 超参数调优

创建网格搜索超参数调优器，添加需要调优的超参数及尝试的数值，这里调优的超参数为：

- 最大迭代次数（maxIter）：20、50 和 100；
- 每一层神经元数量（layers）：[4,5,5,3]和[4,10,10,3]。

```
grid = ParamGridBuilder( ).addGrid(mpc.maxIter,[20,50,100]) \
   .addGrid(mpc.layers,[[4,5,5,3],[4,10,10,3]]).build( )
```

做 3 折交叉验证，得到最佳超参数组合是最大迭代次数（maxIter）为 50 且每一层神经元数量（layers）为[4,10,10,3]。

```
cv = CrossValidator(estimator = mpc, evaluator = MulticlassClassi ficationEvaluator( ), estimatorParamMaps = grid, numFolds = 3)
cvm = cv.fit(iris)
print "maxIter = ", cvm.bestModel._java_obj.parent( ).getMaxIter( )
## maxIter = 50
print "layers = ", cvm.bestModel.layers
## layers = [4,10,10,3]
```

2. 模型评估

调用神经网络分类模型的 transform( )函数预测数值标签，这里为了突出模型，并没有分割训练集和测试集。

```
prediction = cvm.bestModel.transform(iris)
prediction.show(5)
+--------------------+------+----------------------+----------------------+----------+
|        features    |label |    rawPrediction     |       probability    |prediction|
+--------------------+------+----------------------+----------------------+----------+
|[5.1,3.5,1.4,0.2]   |   0.0|[29.0978778584905...|[0.99999999732198...|       0.0|
|[4.9,3.0,1.4,0.2]   |   0.0|[29.1515723469183...|[0.99999999749705...|       0.0|
|[4.7,3.2,1.3,0.2]   |   0.0|[29.2162129772492...|[0.99999999769203...|       0.0|
|[4.6,3.1,1.5,0.2]   |   0.0|[29.0765665736993...|[0.99999999725033...|       0.0|
|[5.0,3.6,1.4,0.2]   |   0.0|[29.0910909254688...|[0.99999999729909...|       0.0|
+--------------------+------+----------------------+----------------------+----------+
```

计算模型精度。

```
accuracy = float(prediction.filter("prediction = label").count( )) /    prediction.count( )
print(accuracy)
## 0.986666666667
```

可以看出，精度远超 50%，有较强的预测能力。

3. 决策边界

画出神经网络模型在 2 个自变量上的决策边界，如图 12-11 所示。

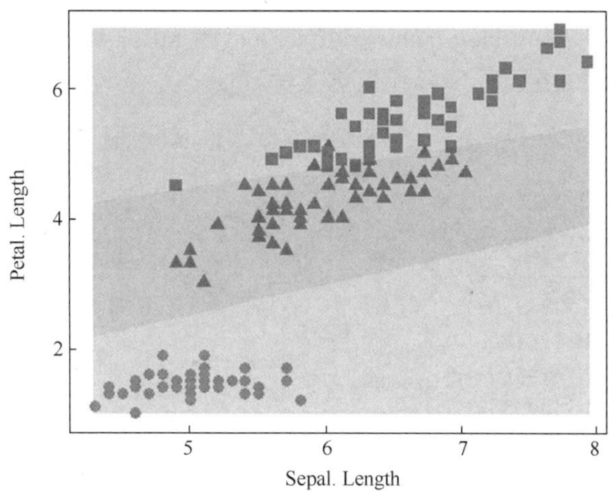

图 12-11　神经网络模型在 2 个自变量上的决策边界（彩色图片见插页）

可以看出，神经网络模型的决策边界可以是一个不规则超曲面。

12.6 协同过滤

协同过滤是一种广泛用于推荐系统的算法。基本思想是填充用户和影片的评分矩阵（即行表示用户、列表示影片、元素表示评分）的缺失值。其中，用户和影片都用多个隐性因子描述，而算法将用户和影片的评分矩阵分解为用户和隐性因子的矩阵以及影片和隐性因子的矩阵，如图 12-12 所示。本节用到的具体算法为交替最小二乘法（alternating least squares）算法。

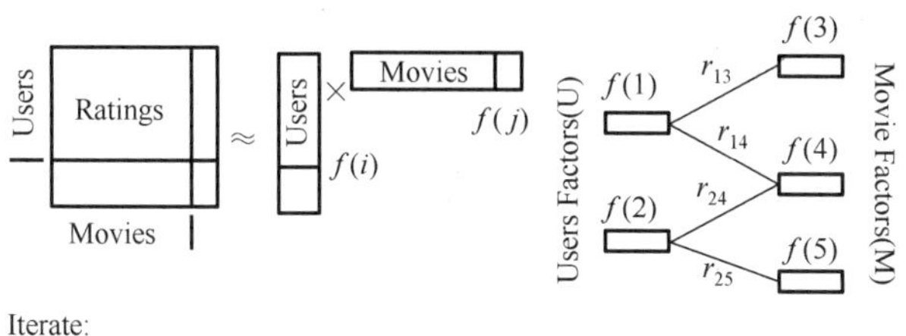

$$f(i) = \arg\min_{w \in \mathbb{R}^d} \sum_{j \in \mathrm{Nbrs}(i)} (r_{ij} - w^{\mathrm{T}} f[i])^2 + \lambda \|w\|_2^2$$

图 12-12　基于矩阵分解的协同过滤算法

本节所有例子基于一个数据集。该数据集是某影评网站中用户对影片的评分数据，包含 1 501 个样本，每个样本包含 4 个属性，如表 12-4 所示

表 12-4　影片评分数据属性定义

属　性	定　义
user_id	用户标识符
movie_id	影片标识符
rating	评分
timestamp	时间戳

该数据集的路径为/root/data/ movie_ratings/movie_ratings.txt。

调用 spark 的 read()函数载入影评数据集 ratings。

```
ratings = spark.read.csv('file:///root/data/movie_ratings/movie_ ratings.txt', header='true', inferSchema='true')
ratings.show(5)
+-----------+----------+------+--------------+
|  user_id  |movie_id|rating| timestamp   |
+-----------+----------+------+--------------+
|          0|         2|     3|1424380312|
|          0|         3|     1|1424380312|
|          0|         5|     2|1424380312|
|          0|         9|     4|1424380312|
|          0|        11|     1|1424380312|
+-----------+----------+------+--------------+
```

只保留预测变量和目标变量，并将用户变量重命名为 user，物品项变量重命名为 item。

```
ratings = ratings.select(ratings.user_id.alias('user'), ratings. movie_id.alias('item'), ratings.rating.cast
('Double'))
ratings.show(5)
+-----+----+------+
|user  |item|rating|
+-----+----+------+
|    0 |   2 |   3.0|
|    0 |   3 |   1.0|
|    0 |   5 |   2.0|
|    0 |   9 |   4.0|
|    0 |  11|   1.0|
+-----+----+------+
```

调用 pyspark.ml.recommendation 程序包中的 ALS( )构造函数创建协同过滤模型，得到超参数列表和每个超参数的默认值。

```
als = ALS( )
print(als.explainParams( ))
alpha: alpha for implicit preference (default: 1.0)
checkpointInterval: set checkpoint interval (>= 1) or disable checkpoint (-1). E.g. 10 means that the cache will get checkpointed every 10 iterations. Note: this setting will be ignored if the checkpoint directory is not set in the SparkContext. (default: 10)
coldStartStrategy: strategy for dealing with unknown or new users/items at prediction time. This may be useful in cross-validation or production scenarios, for handling user/item ids the model has not seen in the training data. Supported values: 'nan', 'drop'. (default: nan)
finalStorageLevel: StorageLevel for ALS model factors. (default: MEMORY_AND_DISK)
implicitPrefs: whether to use implicit preference (default: False)
intermediateStorageLevel: StorageLevel for intermediate datasets. Cannot be 'NONE'. (default: MEMORY_AND_DISK)
itemCol: column name for item ids. Ids must be within the integer value range. (default: item)
maxIter: max number of iterations (>= 0). (default: 10)
nonnegative: whether to use nonnegative constraint for least squares (default: False)
numItemBlocks: number of item blocks (default: 10)
numUserBlocks: number of user blocks (default: 10)
```

```
predictionCol: prediction column name. (default: prediction)
rank: rank of the factorization (default: 10)
ratingCol: column name for ratings (default: rating)
regParam: regularization parameter (>= 0). (default: 0.1)
seed: random seed. (default: 593367982098446717)
userCol: column name for user ids. Ids must be within the integer value range. (default: user)
```

可以看出，有多个可调优的超参数，最常用的包括：

- 参数 regParam：规则化参数；
- 参数 rank：矩阵分解的秩，即隐性因子数量。

1. 超参数调优

创建网格搜索超参数调优器，添加需要调优的超参数及尝试的数值，这里调优的超参数为：

- 规则化参数（regParam）：0.1、0.2 和 0.5；
- 矩阵分解的秩（rank）：5、10 和 20。

```
grid = ParamGridBuilder( ).addGrid(als.regParam,[0.1,0.2,0.5]) \
   .addGrid(als.rank,[5,10,20]).build( )
```

做 3 折交叉验证器，得到最佳超参数组合是规则化参数（regParam）为 0.1 且矩阵分解的秩（rank）为 20。

```
cv = CrossValidator(estimator = als, evaluator = Regression Evaluator(labelCol="rating"),
   estimatorParamMaps = grid, numFolds = 3)
cvm = cv.fit(ratings)
print "regParam = ", cvm.bestModel._java_obj.parent( ).getRegParam( )
## regParam =    0.1
print "rank = ", cvm.bestModel._java_obj.parent( ).getRank( )
## rank =    20
```

2. 模型评估

调用协同过滤模型的 transform()函数预测数值标签，这里为了突出模型，并没有分割训练集和测试集。

```
prediction = cvm.bestModel.transform(ratings)
prediction.show(5)
+----+-----+-------+--------------+
|user |item | rating |  prediction |
+----+-----+-------+--------------+
|   26|   31|      1.0|0.83177257 |
|   27|   31|      1.0| 1.0926105  |
|   12|   31|      4.0| 2.9419067  |
|   13|   31|      1.0| 1.0841627  |
|     5|   31|      1.0| 1.1632977  |
+----+-----+-------+--------------+
```

计算模型性能指标。

```
re = RegressionEvaluator(labelCol = 'rating')
re.evaluate(prediction, {re.metricName:'r2'})
```

```
## 0.9111345033135821
re.evaluate(prediction, {re.metricName:'mse'})
## 0.12518354293604503
```

可以看出，R^2 远超 50%，有较强的预测能力。

小　结

本章主要介绍了在 Spark 中常用的有监督学习模型，包括线性回归、逻辑回归、决策树、随机森林、神经网络和协同过滤模型。每个有监督学习模型又进一步分为回归问题、二分类问题和多分类问题。这些有监督学习模型各有各的适用场景。对于分类问题，不同的模型有不同的决策边界，逻辑回归模型的决策边界为线性，决策树和随机森林模型的决策边界为平行坐标轴，神经网络模型的决策边界可以是非线性的超曲面。

本章使用 Spark 中的 MLlib 实现有监督学习，不同模型的调用格式保持一致。

习　题

1. 各种有监督学习模型的特点是什么？主要从适用场景、变量选择、复杂度和效率等角度叙述。

2. 对于每个有监督学习模型，选取一到两个新的超参数进行调优。

3. 使用不归一化的数据输入神经网络模型，评估结果。

第 13 章 Spark 无监督学习模型

本章主要介绍使用 Spark 中的 pyspark.ml（ML）程序包实现常用的无监督学习模型，如聚类、主成分分析和关联规则挖掘。

无监督学习定义为，训练样本标记信息是未知的或未提供给模型，目标是通过对无标记训练样本的训练来揭示数据的内在性质和规律，为进一步的数据分析提供基础。本章主要介绍聚类模型、主成分分析、关联规则挖掘。

载入本章需要用到的程序包。

```
from pyspark.ml.clustering import KMeans
from pyspark.ml.feature import VectorAssembler, StringIndexer, PCA
from pyspark.ml.classification import DecisionTreeClassifier
from pyspark.sql.functions import split, col, regexp_replace, desc
from pyspark.ml.fpm import FPGrowth
```

本章同样基于鸢尾花卉数据集。数据集的路径为～/data/ iris/iris.csv

该数据集聚类时只输入前 4 个变量，聚类结果可以与物种（变量 species）进行比较，该变量包含了 3 个类别，0 为 Setosa，1 为 Versicolour，2 为 Virginica，每类 50 个样本。

调用 spark 的 read()函数载入鸢尾花卉数据集 iris。

```
iris = spark.read.csv('file:///root/data/iris/iris.csv',
  header='true', inferSchema='true')
iris.show(5)
+------------+-----------+------------+-----------+-------+
|sepal_length|sepal_width|petal_length|petal_width|species|
+------------+-----------+------------+-----------+-------+
|         5.1|        3.5|         1.4|        0.2|      0|
|         4.9|        3.0|         1.4|        0.2|      0|
|         4.7|        3.2|         1.3|        0.2|      0|
|         4.6|        3.1|         1.5|        0.2|      0|
|         5.0|        3.6|         1.4|        0.2|      0|
+------------+-----------+------------+-----------+-------+
```

做特征变量的向量封装。

```
va = VectorAssembler(inputCols=['sepal_length','sepal_width', 'petal_length','petal_width'],
  outputCol = 'features')
iris = va.transform(iris)
```

做目标变量的类别编码。

```
si_species = StringIndexer(inputCol = 'species', outputCol = 'label')
sim_species = si_species.fit(iris)
iris = sim_species.transform(iris)
```

只保留特征变量和目标变量。

```
iris = iris.select(iris.features, iris.label)
iris.show(5)
+-----------------+------+
|      features   |label |
+-----------------+------+
|[5.1,3.5,1.4,0.2]|   0.0|
|[4.9,3.0,1.4,0.2]|   0.0|
|[4.7,3.2,1.3,0.2]|   0.0|
|[4.6,3.1,1.5,0.2]|   0.0|
|[5.0,3.6,1.4,0.2]|   0.0|
+-----------------+------+
```

下面将依次介绍常用的无监督学习模型，重点会覆盖：

- 模型在聚类和主成分分析中的应用；
- 模型在 pyspark.ml 程序包中的实现方法；
- 模型各超参数的含义和影响。

13.1　k 均值聚类模型

给定样本集 $D=\{\boldsymbol{x}_1,\boldsymbol{x}_2,\cdots,\boldsymbol{x}_m\}$，k 均值（$k$-means）算法针对聚类所得划分 $C=\{C_1,C_2,\cdots,C_k\}$ 最小化平方误差为：

$$E=\sum_{i=1}^{k}\sum_{\boldsymbol{x}\in C_i}||\boldsymbol{x}-\boldsymbol{\mu}_i||_2^2$$

式中，$\boldsymbol{\mu}_i=\frac{1}{|C_i|}\sum_{\boldsymbol{x}\in C_i}\boldsymbol{x}$ 是簇 C_i 的均值向量。E 在一定程度上刻画了簇内样本围绕簇均值向量的紧密程度，E 越小则簇内样本相似度越高。

最小化 E 并不容易，找到它的最优解需要考察样本集 D 所有可能的簇划分，这是一个 NP 难问题。因此，k 均值算法采用了贪心策略，通过迭代优化来近似求解，算法流程如下所示。

输入： 样本集 $D=\{\boldsymbol{x}_1,\boldsymbol{x}_2,\cdots,\boldsymbol{x}_m\}$

　　聚类簇数 k

过程： 函数 TreeGenerate(D, A)

1: 从 D 中随机选择 k 个样本作为初始均值向量 $\{\boldsymbol{\mu}_1,\boldsymbol{\mu}_2,\cdots,\boldsymbol{\mu}_k\}$

2: **repeat**

3: 　　令 $C_i=\varnothing(1\leqslant i\leqslant k)$

4: 　　**for** $j=1,2,\cdots,m$ **do**

5: 计算样本 $\boldsymbol{x}_j$ 与各均值向量 $\boldsymbol{\mu}_i (1 \leqslant i \leqslant k)$ 的距离： $d_{ji} = \|\boldsymbol{x}_j - \boldsymbol{\mu}_i\|_2$

6: 根据距离最近的均值向量确定 $\boldsymbol{x}_j$ 的簇标记： $\lambda_j = \mathrm{argmin}_{i \in \{1,2,\cdots,k\}} d_{ji}$

7: 将样本 $\boldsymbol{x}_j$ 划入相应的簇： $C_{\lambda_j} = C_{\lambda_j} \cup \{\boldsymbol{x}_j\}$

8: **end for**

9: **for** $i = 1,2,\cdots,k$ **do**

10: 计算新均值向量： $\boldsymbol{\mu}_i' = \frac{1}{|C_i|}\sum_{\boldsymbol{x} \in C_i} \boldsymbol{x}$

11: **if** $\boldsymbol{\mu}_i' \neq \boldsymbol{\mu}_i$ **then**

12: 将当前均值向量 $\boldsymbol{\mu}_i$ 更新为 $\boldsymbol{\mu}_i'$

13: **else**

14: 保持当前均值向量不变

15: **end if**

16: **end for**

17: **until** 当前均值向量均未更新

输出： 簇划分 $C = \{C_1, C_2, \cdots, C_k\}$

以下是一个 k 均值模型（ $k = 2$ ）的全过程，样本集中有 100 个样本，颜色表示簇标记，最大的表示均值向量。如图 13-1 所示，整个过程中，2 个图为一组表示一次迭代，每一组中：

- 左图：根据距离最近的均值向量确定每个样本的簇标记；
- 右图：计算新均值向量。

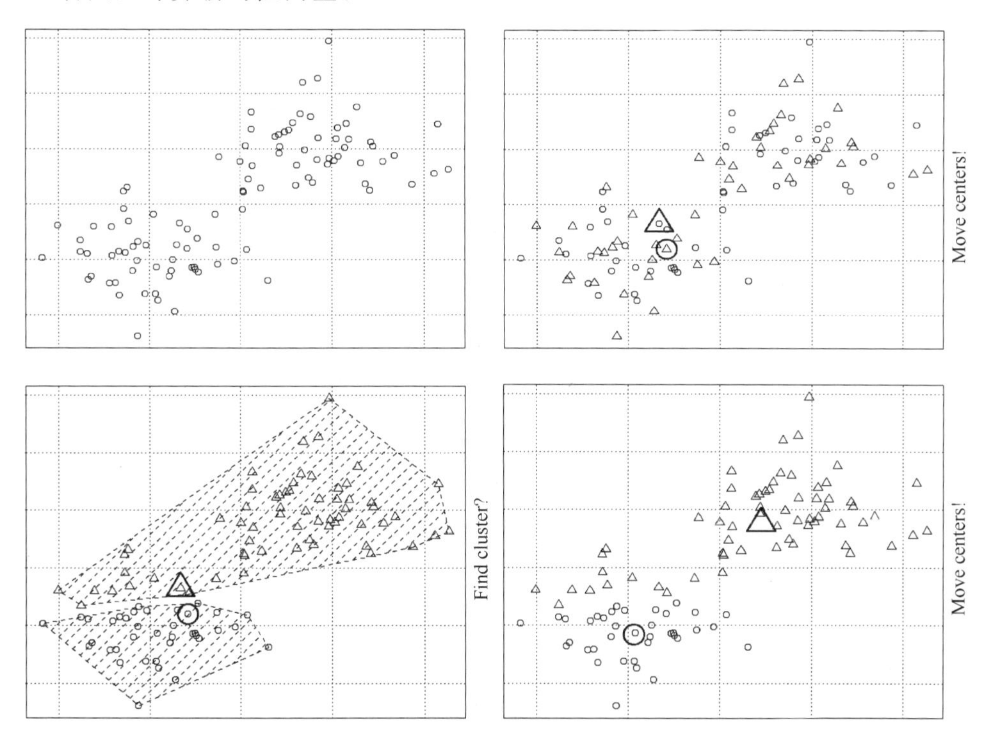

图 13-1 k 均值聚类模型（ $k = 2$ ）的全过程

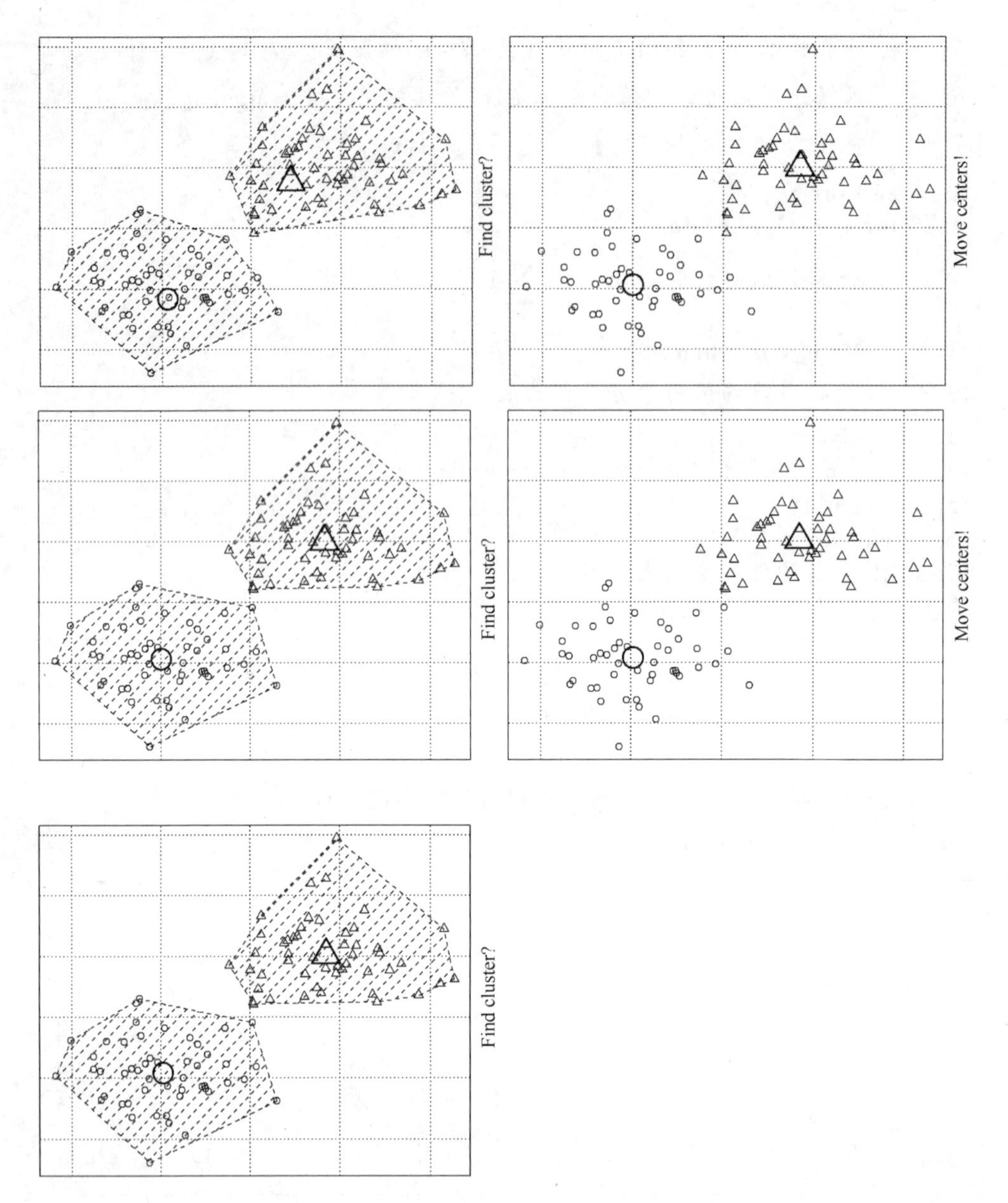

图 13-1　k 均值聚类模型（ $k=2$ ）的全过程（续）

调用 pyspark.ml.clustering 程序包中的 KMeans()构造函数创建 k 均值算法，其中：

- 参数 k 表示聚类得到簇的数量；
- 参数 seed 表示随机数种子。

调用 k 均值算法的 explainParam( )函数得到超参数列表和每个超参数的默认值和当前值。

```
km = KMeans(k = 3, seed = 123)
print(km.explainParams( ))
featuresCol: features column name. (default: features)
```

```
initMode: the initialization algorithm. This can be either "random" to choose random points as initial cluster centers, or "k-means||" to use a parallel variant of k-means++ (default: k-means||)
initSteps: steps for k-means initialization mode (default: 5)
k: number of clusters to create (default: 2, current: 3)
maxIter: max number of iterations (>= 0). (default: 20)
predictionCol: prediction column name. (default: prediction)
seed: random seed. (default: -7649703878154674547, current: 123)
tol: the convergence tolerance for iterative algorithms. (default: 0.0001)
```

1. 模型训练

调用 k 均值算法的 fit()函数做训练，输入参数为数据框，返回训练后的 k 均值模型。

```
kmm = km.fit(iris)
```

调用 k 均值模型的 clusterCenters()函数得到簇的中心。

```
print(kmm.clusterCenters())
## [array([ 5.88360656, 2.74098361, 4.38852459, 1.43442623]), array([ 5.006, 3.418, 1.464, 0.244]), array([ 6.85384615, 3.07692308, 5.71538462, 2.05384615])]
```

调用 k 均值模型的 transform()函数预测簇的标签。

```
prediction = kmm.transform(iris)
prediction.show(5)
+-----------------+-----+----------+
|     features    |label|prediction|
+-----------------+-----+----------+
|[5.1,3.5,1.4,0.2]|  0.0|         1|
|[4.9,3.0,1.4,0.2]|  0.0|         1|
|[4.7,3.2,1.3,0.2]|  0.0|         1|
|[4.6,3.1,1.5,0.2]|  0.0|         1|
|[5.0,3.6,1.4,0.2]|  0.0|         1|
+-----------------+-----+----------+
```

2. 模型评估

对于聚类问题，可以利用已知的样本类别标签评估聚类模型。

调用数据框的 crosstab()函数得到聚类结果和实际物种的列联表。

```
prediction.crosstab('label', 'prediction').show()
+--------------------+---+---+---+
| label_prediction   |  1|  0|  2|
+--------------------+---+---+---+
|                 2.0|  0| 48|  2|
|                 1.0|  0| 14| 36|
|                 0.0| 50|  0|  0|
+--------------------+---+---+---+
```

可以看出，类别序号 0 代表的 setosa 完全对应 1 个簇，而类别序号 1 代表的 versicolor 和类别序号 2 代表的 virginica 基本也分成了 2 个簇。

13.2 主成分分析模型

主成分分析（Principal Component Analysis，PCA）是一种统计方法。通过正交变换将一组可能存在相关性的变量转换为一组线性不相关的变量，转换后的这组变量叫主成分。

调用 pyspark.ml.feature 程序包中的 PCA()构造函数创建主成分分析算法，其中：

- 第 1 个参数 k 表示保留主成分的数量；
- 第 2 个参数 inputCol 表示输入向量的变量名称；
- 第 3 个参数 outputCol 表示输出主成分的变量名称；

调用主成分分析算法的 explainParam( )函数得到超参数列表和每个超参数的默认值和当前值。

```
pca = PCA(k = 2, inputCol = 'features', outputCol = 'pca_features')
print(pca.explainParams( ))
inputCol: input column name. (current: features)
k: the number of principal components (current: 2)
outputCol: output column name. (default: PCA_414a936a7842d3823547__output, current: pca_features)
```

1. 模型训练

调用主成分分析器的 fit()函数做训练，输入参数为数据框，返回训练后的主成分分析模型。

```
pcam = pca.fit(iris)
```

调用主成分分析模型的 transform()函数得到前 2 个主成分。

```
pca_features = pcam.transform(iris)
pca_features.show(5)
+-----------------+-------+------------------------+
|     features    |species|       pca_features     |
+-----------------+-------+------------------------+
| [5.1,3.5,1.4,0.2]|      0| [-2.8271359726789... |
| [4.9,3.0,1.4,0.2]|      0| [-2.7959524821488... |
| [4.7,3.2,1.3,0.2]|      0| [-2.6215235581650... |
| [4.6,3.1,1.5,0.2]|      0| [-2.7649059004742... |
| [5.0,3.6,1.4,0.2]|      0| [-2.7827501159516... |
+-----------------+-------+------------------------+
```

2. 比较主成分特征和原始特征建模性能

输入原始特征建立决策树模型，其中超参数最大深度（maxDepth）为 3 且最小信息增益（minInfoGain）为 0.001，计算模型的精度。

```
dtc = DecisionTreeClassifier(maxDepth = 3, minInfoGain = 0.001)
dtcm = dtc.fit(iris)
prediction = dtcm.transform(iris)
accuracy = float(prediction.filter("prediction = label").count( )) /   prediction.count( )
```

```
print(accuracy)
# 0.973333333333
```

输入主成分特征建立决策树模型，其中超参数与上一步保持一致，计算模型的精度。

```
dtc = DecisionTreeClassifier(maxDepth = 3, minInfoGain = 0.001, featuresCol = 'pca_features')
dtcm = dtc.fit(pca_features)
prediction = dtcm.transform(pca_features)
accuracy = float(prediction.filter("prediction = label").count()) /   prediction.count()
print(accuracy)
# 0.946666666667
```

可以看出，相比于原始的 4 个特征，前 2 个主成分建立的模型性能相差不多，但减少了一半输入数据量。

13.3 关联分析模型

关联规则用于表达项集之间的联系，形式如同 $A \Rightarrow B$，其中 A 和 B 是两个不相交的项集，分别称为规则的左侧项集（left-hand side）和右侧项集（right-hand side）。关联规则有 3 个最重要的衡量指标，分别为

- 支持度（support）：样本中同时包含 A 和 B 的样本比例，定义为

$$\text{support}(A \Rightarrow B) = P(A \cup B)$$

- 置信度（confidence）：包含 A 的样本中同时包含 B 的样本比例，定义为

$$\text{confidence}(A \Rightarrow B) = P(B \mid A) = \frac{P(A \cup B)}{P(A)}$$

- 提升度（lift）：置信度与包含 B 的样本比例之比，定义为

$$\text{lift}(A \Rightarrow B) = \frac{\text{confidence}(A \Rightarrow B)}{P(B)} = \frac{P(A \cup B)}{P(A)P(B)}$$

挖掘关联规则的经典算法是 FP-growth 算法，首先计算并找出频繁项，再用一种树结构编码每个交易，并从中得到频繁项。

1. 案例数据

本节的所有例子基于某杂货店一个月的交易数据，包含 9 835 条交易记录，涵盖了 169 个商品类别，每一行代表一笔交易，包含了这笔交易所涉及的商品类别，用逗号分隔。该数据集的路径为/root/data/Groceries/Groceries.csv。

调用 spark 的 read()函数载入交易数据集 Groceries。

调用 regexp_replace()函数去除多余的字符串{}，调用 split()函数按逗号分隔符分隔每条交易记录。

```
Groceries = spark.read.csv('file:///root/data/Groceries/Groceries.csv', header='true', inferSchema='true')
Groceries = Groceries.withColumn("items", split(regexp_replace(col("items"), "\{|\}", ""), ","))
Groceries.show(5, truncate=False)
```

```
+--------------------------------------------------------------------------------+
|items                                                                           |
+--------------------------------------------------------------------------------+
|[citrus fruit, semi-finished bread, margarine, ready soups]                     |
|[tropical fruit, yogurt, coffee]                                                |
|[whole milk]                                                                    |
|[pip fruit, yogurt, cream cheese , meat spreads]                                |
|[other vegetables, whole milk, condensed milk, long life bakery product]        |
+--------------------------------------------------------------------------------+
```

2. 关联规则挖掘

调用 pyspark.ml.fpm 程序包中的 FPGrowth()构造函数创建 FP-growth 算法，其中：

- 第 1 个参数 minSupport 表示最小支持度阈值，默认为 0.3；
- 第 2 个参数 minConfidence 表示最小置信度阈值，默认为 0.8；
- 第 3 个参数 itemsCol 表示项的列表的变量名称，默认为 items；
- 第 4 个参数 itemsCol 表示预测值的变量名称，默认为 prediction。

调用FP-growth算法的explainParam()函数得到超参数列表和每个超参数的默认值和当前值。

```
fpg = FPGrowth(minSupport=0.01, minConfidence=0.5)
print(fpg.explainParams( ))
itemsCol: items column name (default: items)
minConfidence: Minimal confidence for generating Association Rule. [0.0, 1.0]. minConfidence will not affect the mining for frequent itemsets, but will affect the association rules generation. (default: 0.8, current: 0.5)
minSupport: Minimal support level of the frequent pattern. [0.0, 1.0]. Any pattern that appears more than (minSupport * size-of-the-dataset) times will be output in the frequent itemsets. (default: 0.3, current: 0.01)
numPartitions: Number of partitions (at least 1) used by parallel FP-growth. By default the param is not set, and partition number of the input dataset is used. (undefined)
predictionCol: prediction column name. (default: prediction)
```

模型训练如下：

调用 FP-growth 的 fit()函数做训练，输入参数为数据框，返回训练后的 FP-growth 模型。

```
fpgm = fpg.fit(Groceries)
```

使用 FP-growth 模型的 freqItemsets 属性得到项集频次数据框，其中：

- 变量 items 表示项集；
- 变量 freq 表示频次。

```
fpgm.freqItemsets.show(5, truncate=False)
+------------------------------------------------+----+
|                     items                      |freq|
+------------------------------------------------+----+
| [canned vegetables]                            |106 |
| [pork]                                         |567 |
```

```
| [pork, rolls/buns]                              |111 |
| [pork, other vegetables]                        |213 |
| [pork, other vegetables, whole milk]            |100 |
+-------------------------------------------------+----+
```

调用数据框的 sort()函数按频次降序排列，并查看前 10 个频繁项集。

```
fpgm.freqItemsets.sort(desc("freq")).show(10, truncate=False)
+-----------------------+-----+
|        items          | freq |
+-----------------------+-----+
|[whole milk]           |2513|
|[other vegetables]     |1903|
|[rolls/buns]           |1809|
|[soda]                 |1715|
|[yogurt]               |1372|
|[bottled water]        |1087|
|[root vegetables]      |1072|
|[tropical fruit]       |1032|
|[shopping bags]        | 969 |
|[sausage]              | 924 |
+-----------------------+-----+
```

使用 FP-growth 模型的 associationRules 属性得到关联规则数据框，其中：

- 变量 antecedent 表示左边项集；
- 变量 consequent 表示右边项集；
- 变量 confidence 表示置信度；
- 变量 lift 表示提升度。

```
fpgm.associationRules.show(5, truncate=False)
+---------------------------------+--------------------+----------------------+-----------------------+
|          antecedent             |    consequent      |     confidence       |          lift         |
+---------------------------------+--------------------+----------------------+-----------------------+
|[root vegetables, yogurt]        |[other vegetables]  |0.5                   |2.5840777719390435|
|[root vegetables, yogurt]        |[whole milk]        |0.562992125984252     |2.2033535849801504|
|[tropical fruit, yogurt]         |[whole milk]        |0.5173611111111112|2.0247698081089447|
|[whipped/sour cream, yogurt]     |[whole milk]        |0.5245098039215687|2.052747282757114  |
|[pip fruit, other vegetables]    |[whole milk]        |0.5175097276264592|2.0253514409893456|
+---------------------------------+--------------------+----------------------+-----------------------+
```

调用数据框的函数 sort()按置信度降序排列，并查看前 10 条规则。

```
fpgm.associationRules.sort(desc("confidence")).show(10, truncate=False)
+-----------------------------------+--------------------+----------------------+---------------------------+
|             antecedent            |    consequent      |     confidence       |            lift           |
+-----------------------------------+--------------------+----------------------+---------------------------+
|[citrus fruit, root vegetables]    |[other vegetables]|0.5862068965517241|3.0296084222733612   |
|[tropical fruit, root vegetables]  |[other vegetables]|0.5845410628019324|3.020999134344196    |
|[curd, yogurt]                     |[whole milk]       |0.5823529411764706|2.27912502048173     |
```

```
|[butter, other vegetables]          |[whole milk]       |0.5736040609137056|2.244884973770909        |
|[tropical fruit, root vegetables]   |[whole milk]       |0.5700483091787439|2.2309690094599866       |
|[root vegetables, yogurt]           |[whole milk]       |0.562992125984252 |2.2033535849801504       |
|[domestic eggs, other vegetables]   |[whole milk]       |0.5525114155251142|2.1623357627097084       |
|[whipped/sour cream, yogurt]        |[whole milk]       |0.5245098039215687|2.052747282757114        |
|[root vegetables, rolls/buns]       |[whole milk]       |0.5230125523012552|2.04688756541299         |
|[pip fruit, other vegetables]       |[whole milk]       |0.5175097276264592|2.0253514409893456       |
+------------------------------------+-------------------+------------------+-------------------------+
```

小　　结

本章主要介绍了在 Spark 中常用的无监督学习模型，包括 k 均值聚类、主成分分析和关联分析模型。k 均值聚类需要事先指定簇的数量，基于样本距离做聚类，且会因为初始点选取的不同而得到不同的聚类结果。主成分分析通过正交变换将一组可能存在相关性的变量转换为一组线性不相关的主成分。关联分析模型主要介绍了 FP-Growth 算法挖掘关联规则。

本章使用 Spark 中的 MLlib 实现无监督学习。

习　　题

1. 各种无监督学习模型的特点是什么？主要从适用场景、输入参数、复杂度和效率等角度叙述。

2. 调整 k 均值聚类模型的参数 k，对比结果的不同。

3. 调整 FP-Growth 关联规则模型算法的支持度、置信度和提升度，对比结果的不同。